THE STRAWBERRY CONNECTION

STRAWBERRY COOKERY

with flavour, fact and folklore,
from memories, libraries and kitchens
of old and new friends – and strangers

compiled at
cranberrie cottage in
granville centre, nova scotia, by

BEATRICE ROSS BUSZEK

The recipes in this book
were hand-lettered by
Beatrice Ross Buszek
at Cranberrie Cottage
in Nova Scotia, Canada

Published by Nimbus Publishing Limited
        P.O. Box 9301, Station A
        Halifax, Nova Scotia
        B3K 5N5

ISBN 0-920852-31-9

First printing 1984
Second printing 1986
Third printing 1988
Printed and bound in Canada by Wm. MacNab & Son Ltd., Halifax, Nova Scotia

Distributed in the United States by Yankee Books, Depot Square, Peterborough,
N.H. 03458

*This book is dedicated
to
Strawberry Specialists,
Growers and Lovers
with
special tribute
to
Nova Scotia Strawberry Pioneers
who have sweetened our days with
the fruits of their curiosity,
initiative and patience.*

Acknowledgement

The first and greatest acknowledgement goes to Dr. Stephen Wilhelm and James E. Sagen of the Department of Pomology, Division of Agricultural and Natural Resources, University of California at Berkeley. Their book, **The History of the Strawberry from Ancient Gardens to Modern Markets** was fascinating and very helpful in the preparation of **The Strawberry Connection.**

I hope that Dr. Wilhelm will not be affronted with this anthropomorphic approach to the strawberry story. It is a reflection of his scholarship that the research fired my imagination and led me into the complex genetics of the modern strawberry and its fated beginning. It is, admittedly, entirely out of my professional field and hardly the stuff of which cookbooks are made. On the other hand, that is the challenge, to make the connection between the past and the present, and between the kitchen and nature's gifts. Throughout the strawberry book are quotes or ideas and reprints of old prints, most of which are attributed to Dr. Wilhelm's research.

Acknowledgement is also made to all those, living and dead, who wrote the books and articles listed on page 213. Fletcher's work was especially helpful as was Hedrick's 1925 **Small Fruits of New York**, a gift to me from the Pomology Department at Cornell University in Ithaca. The American and Canadian Departments of Agriculture were very helpful and Horticulturist Robert Murray from the N.S. Department of Agriculture brought me a huge box of literature, research, statistics and recipes. General Foods Consumer Centre in Ontario sent me an array of tested recipes, especially jams and jellies and a photo of strawberries and rhubarb. All these were most appreciated.

My daughter, Christine McEntee, did the artwork for the cover and some of the sketches. She took the time from a first baby (my fourth grandchild, Thomas Joseph) and came to my aid. And there were so many others who helped, one way or another. To all those unnamed here, I acknowledge with gratitude.

STRAWBERRIES delightsome to the tast

acceptable to the stomach

Muffins

Toppings

Jams & Jellies

Cakes & Cookies

exciting to the appetite

Potpourri

quenching to the thirst

Soups

Drinks

excellently cooling

> ### STRAWBERRIES
> *"delightsome to the tast,*
> *acceptable to the stomack,*
> *exciting to the appetite,*
> *quenching to the thirst*
> *and excellently cooling."*
> *(1638)*

INTRODUCTION

The Strawberry Connection, unlike its Cranberry, Blueberry and Sugar Bush siblings, was conceived thousands of miles away from my home at Cranberrie Cottage in the Annapolis Valley of Nova Scotia.

There is a similarity between the seemingly fated beginnings of all the **Connections**, and the history of the modern day strawberry. In the case of **The Strawberry Connection**, the comparison is the more striking as the book is uniquely the result of a sudden coming together of a series of unexpected events in time and place. In much the same way, the modern strawberry is viewed by horticulturists as a "mere accident of fate", a chance hybridization in Europe of two wild strawberry plants from North and South America, brought there nearly two centuries earlier.

After living outside of Canada for many years, I returned to Nova Scotia and bought a little house in the country with eight acres of land, one of which was covered with cranberries. The subsequent bog adventure culminated in **The Cranberry Connection**, published in the fall of 1977. Almost immediately, book dealers, cooks, friends, neighbours and berry specialists began to query about the next book, most of them even then, urging me towards strawberry fields. I backed off. Wasn't it presumptuous to try to improve on the ultimate - fresh strawberries with cream? Nor could I envision enough variation on that theme to make a book. I dismissed the idea. Instead, one day, across from the house, I found myself surrounded by dewy downy bunches of black and blue. The years faded and I was a little girl, off to pick berries in the pasture. This encounter with my Maritime roots led to **The Blueberry Connection**. Then it was March, 1982. I had left the berries in the bogs and the pastures, and was off, in the mud and patches of snow, to the *Sucrérie*, absorbed in the mystery of the Sugar Maple miracle. When **The Sugar Bush Connection** was published I was already exploring other spheres and unrelated matters, having put the box marked "Strawberry" underneath the cot in the upstairs workroom. Perhaps it wasn't the right time, but whatever the real reason, it seemed that I was immune to the strawberry frenzy.

Nevertheless, the seeds of **The Strawberry Connection** continued to be sown. More and more people extolled the virtues of this berry/fruit widely distributed all over the world and offering a faster return on investment than any other fruit crop. At a more personal level, it was assumed that "everybody loves strawberries." But the seeds still fell on inhospitable soil. It took an abrupt departure from my life at Cranberrie Cottage, and eight angular months on a family mission in Big Sky country, before those dormant seeds began to germinate. When they did, when the sprouts began to appear, I was transported into a whole new world, a world about which I knew nothing even though I picked lots of wild berries as a girl, and have loved Strawberries with Cream, and Strawberry Shortcake, all of my life. The strawberry story, a fated tale of star-crossed lovers - of a sort - opened me to the deferred adventure, took me down into strawberry fields, and became the brightest caper of all.

The frustrations were legion those years I was writer, publisher, book-keeper and distributor for the **Connections**. Still it was a healthy diversion, both the stimulating challenge of new ideas and all the good people I might otherwise never have met. The greatest frustration was that I wanted to tell the readers everything I had learned about the berries/syrup/fruit, as well as invite them to scrumptious culinary experiences. As a compromise, bits of fact and folklore were inserted between the recipes. In this book I include additional vignettes from the Scarlet past of the modern day strawberry. Few people, if any except the horticulturists and pomologists, know anything about the strawberry story and yet this fruit has touched all of our lives.

The wild strawberry plant grows all over the world from the arctic to the tropics, adapting and surviving with characteristics unique to the region. In ancient times it was taken from the woods and fields and tenderly cared for in private and apothecary gardens. Virgil said the wild berries were "food for a Golden Age", an imaginary time when life was pure and joyous. In medieval times the strawberry was depicted as a religious symbol representing "eternal righteousness". The trifoliate leaves may have been perceived as symbolic of the Holy Trinity. In one painting St. Joseph holds strawberries out to the Christ Child and in another the Angels are gathering the berries for the Virgin Mary and the Child Jesus. The presence of the strawberry in paintings and illuminated manuscripts denoted eternal salvation. Even in its wild early years the strawberry had a varied and fascinating life.

Gardeners and botanists continued to select the best of the wild strawberry plants and cultivated them as varieties. They were unaware of the limitations of cultivation or the importance of the sex of the plants; that there are male, female and hermaphrodite forms of the strawberry. This state of strawberry innocence, even through centuries of fruiting and non-fruiting plants, persisted into the 18th century.

There are many documents that date the wild strawberry to antiquity but the history of the modern strawberry began when the New World was discovered.

Jacques Cartier found the bright red wild berry along the shores of the St. Lawrence and along the Atlantic coast of what is now Canada. He took some plants back to France where they were known as the Scarlet Strawberry of Canada, *Fragaria canadana* or *Fragaria americana*. In the early 1600's explorers and colonists in Virginia found an abundance of fruiting plants among which were "Straberies ... as good and as great as those we have in our English gardens." The seeds, and later the plants, of the wild Virginian strawberry were taken to England and called *Fragaria virginiana*. The strawberry plants from North America were soon cultivated in botanical and medicinal gardens now established in most of the large centres of Europe. Eventually, despite their distinct horticultural differences, it was recognized that the Scarlet Strawberry of Canada and the wild plant from Virginia were one and the same, botanically identical. Once the identity was established the plant was given the name *Fragaria virginiana*. Nearly two centuries would lapse before this strawberry from North America would sire the modern day strawberry.

In 1714 the French government sent a man to South America on military reconnaissance. While there he visited Chile and saw huge wild Chilean strawberries "as big as a hen's egg", being cultivated by the Indians. He took plants back to France but only two survived the long voyage. He would not have understood but Fate was smiling. The survivor strawberry plants were female. The gardeners and botanists, however, were disappointed and perplexed. Instead of the huge Chilean berries they expected, the transplanted plants were barren.

At the Botanical Gardens of Brest in Brittany in 1740, an ornamental planting system was devised whereby the Chilean and Virginian strawberry plants were placed in alternate rows. The flowering runners of the strawberry plants strewed or strawed - or strayed - over the trough and up the hill between the rows, intermingling with each other. A few years later a new race of strawberry appeared in the gardens and the markets of Europe. Nobody could explain where it had come from but recognized it as being vastly superior in all ways to all other varieties. One horticulturist likened the new strawberry to "a foreign maiden from an unknown land". With a bit of imagination and a smile, one could call it a "bundling berry" taking the cue from a well-known early colonial custom in New England that now and then produced bundling babies. Brest became the centre of commercial strawberry production in Europe for the next hundred years.

In 1766 a young gardener at Versailles made what then seemed to be an absurd statement. He suggested that the new plant was an accidental hybrid of the two wild plants from North and South America. His theory was validated by research but old ideas and old ways change slowly. As late as 1911 the famous American botanist, Liberty Hyde Bailey, still did not accept the theory. Eventually, however, *Fragaria chiloensis* was formally wedded to *Fragaria virginiana*, legitimizing the already accidental fruit of their union. Now, at last, through a quirk of fate, the horticultural potential of the strawberry and the importance of its sexuality was recognized, thus beginning a new era of scientific inquiry into plant genetics and physiology. Antoine Nicholas Duchesne, the young gardener from Versailles, is credited with being "the first to recognize that a new race of fruit could arise by accidental crossing of two species." The hybrid offspring of the North and South American fruit became the modern day garden strawberry, the Pine berry, *Fragaria ananas*. When the first plants came to America they were cultivated in a medicinal/botanical garden located at the present site of the Rockefeller Center in New York City. With this new race of strawberry began a "modern era of commercial strawberry culture."

There is still controversy about how the strawberry got its name. The modern spelling dates from the 15th century. One theory dates to 900 A.D. and suggests that the strawberry was originally called "hayberry" as it ripens during haying time. That idea is dismissed by botanists. Another version relates the name to the placing of straw around plants for protection. This is still a popular view but in fact the name was used before this 19th century practice. Early in the 13th century children gathered the wild berries and sold them on the streets. The berries were threaded on straws and the cry was, "Who will buy my straw of berries", or, "A penny for a straw of berries." This is my favourite version. An 1820 dictionary contends that the name derived from the character of the strawberry, "... for it is a plant, the running stems of which are strewed (strawed, anciently used) over the ground." Few of the dictionaries or encyclopedia agree on the meaning of the word. The name is very old and the exact derivation probably never will be known. Romantic, homely, colloquial meanings tend to be lost or altered over the centuries.

The Strawberry Connection goes to press in a few days. In the eight months since returning from New Mexico I have reread stacks of notes and made many more, and tried all sorts of ideas using strawberries. Only one recipe using Jello got into the book but many recipes use wine, liqueur and herbs. Such an affinity dates back hundreds of years. The over two hundred concoctions in **The Strawberry Connection** are arranged as in the earlier books, and in-between are bits of scarlet strawberry trivia, an inducement to whet the appetite for this ubiquitous berry that has sweetened our tables and lives since antiquity.

The lack of traditional cookbook order ... is not by chance. The design, or lack of same, is a sort of outpouring of recipes, fact and folklore. My mother has an old scribbler with the same peculiar kind of order, and in it, either written, printed or pasted, are the recipes of her life. I always marvel how she can find a certain recipe, almost as if she knows where each one fits in the life and thickness of the scribbler. Her system introduced me to concoctions I would never have known had I relied only on an index or if, for example, all the pies and only pies were arranged together. I compromised ... providing a table of contents to balance the outpouring of recipes.

Not everyone can go down to strawberry fields, but the Cambridge Fellow was right when he said in 1548 that "Everyman knoweth wel inough where strawberries growe", because strawberries grow everywhere - from Kentville to Kashmir to Kyoto. Here at Cranberrie Cottage the wild strawberries grow in the field where the blueberries grow, and across the highway along the dirt road towards the mountain, and along the edge of the wooded west lot line. In the back yard there are plants around the lilacs, the forsythia and the japonica. Seeing them gives me a sense of continuity with the past. Strawberries are, indeed, the "fruit of fruits" and worthy as "food for a Golden Age" - or for any age. **Strawberry Fields Forever!**

Cranberrie Cottage
July, 1984

Fragaria virginiana

This is the plant that Jacques Cartier
found in Canada. The early English
colonists later found the same plant
in Virginia. It found a mate in Europe
and became one of the parents of
today's Strawberry.

1

"There are plenty of gooseberries, strawberries and
roses ... the lands where there
are no woods are full of peas,
gooseberries, white and red
strawberries, raspberries ...
(along shores of the St. Lawrence)
        Discovery of St. Lawrence
                    River of Canada

SCOTTISH STRAWBERRY SQUARES

1 cup flour — 1 egg — ½ cup butter
1 tsp. baking powder — 1 T. milk
Dash of salt   — Grated lemon peel

Mix all except lemon peel. Spread
in square baking pan. Cover
liberally with strawberry
jam. For this recipe, "boughten
jam is best". Topping —
Mix 1 egg with 1 cup vanilla
sugar, 2 cups coconut and
2 Tablespoons melted butter.
Bake in moderate oven about
30-40 minutes. Sprinkle grated
lemon peel over the top.

2

Some advice for the gardener:

"Do not make the strawberry bed too rich; a gravelly loam is the best soil; guano and woods-mold the best manure. An abundance of water during the fruiting season is indispensable. Aim to grow berries not vines. Set plants in rows, two feet apart and one foot between plants. Keep the runners cut close and let the old plants grow while they bear well, and then cut them out leaving a new root to grow in its place. Cover the bed with leaves in the fall and don't rake them off in the spring; they keep the ground free of weeds and fruit clean."

BRANDIED FRUIT TOPPING

Beat whipping cream until it is thick. Fold into the cream, fresh sliced straw-berries that have stood at least six hours in brandy. If you must sweeten the topping, be stingy with the sugar.

The strawberry is an aggregate fruit. It is made up of many enlarged ovaries massed on a single stem. The growth of the fruit is controlled by seeds and there is no fruit unless the seeds are fertilized. Pollen is a necessity for strawberry growth and development. Bees are used extensively throughout the world, to pollinize the plants. (Each "seed" of strawberry is in fact a fruit.)

SWEET SOUP

2 cups strawberry juice
2¼ cups cold water
4 T. cornstarch — 1 cup sugar
1 cup whipped cream

Add juice and sugar to boiling water. Mix cornstarch in about 3-4 T. water. Combine the two mixtures. Serve cold with whipped cream.
(This recipe makes a fine topping over a rice pudding)

4

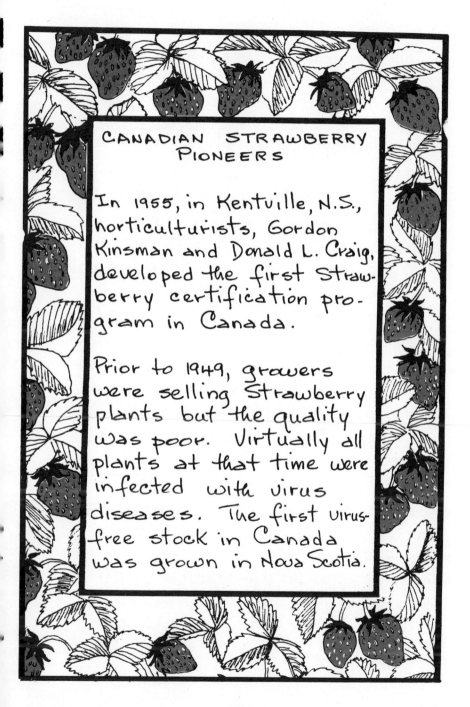

CANADIAN STRAWBERRY PIONEERS

In 1955, in Kentville, N.S., horticulturists, Gordon Kinsman and Donald L. Craig, developed the first strawberry certification program in Canada.

Prior to 1949, growers were selling strawberry plants but the quality was poor. Virtually all plants at that time were infected with virus diseases. The first virus-free stock in Canada was grown in Nova Scotia.

"The learned doctors say that strawberries are cold and moist in the
third degree; the properties of the
fruit are also present throughout
the plant." (1485)

STRAWBERRY MUFFINS

1 cup small Strawberries
1/4 cup butter   - 1/3 cup sugar
2 2/3 cup flour - 4 tsp. baking powder
1 cup milk   - 1/2 tsp. salt
1 egg, beaten

Cream butter and sugar, add
well-beaten egg. Sift flour,
salt and baking powder,
Keeping aside about 1/4 cup
flour. Mix all together, adding
milk slowly. Toss straw-
berries in 1/4 cup flour.
Fold into the mixture. Bake
in a hot oven about 15-20
minutes.

"cold and moist in the third
degree" is early medical Term-
inology. Author of quote is
unknown but words are
recorded.

The Strawberry is the most widely distributed economic crop in the world. Strawberries are usually graded when picked.

COUNTRY PUNCH

Simmer together <u>4 cups straw-berries</u> and <u>5 cups Rhubarb,</u> cut but not peeled. Be very stingy with water - barely cover bottom of pan. When mushy, strain. Measure liquid into an enamel Kettle. Add <u>1/3 cup Sugar for each 1 cup juice.</u> Stir until Sugar is dissolved. Cool. Add strained juice of <u>5 oranges</u> and <u>3 lemons</u>. Chill. When ready to serve, add <u>1 quart dry chilled Gingerale.</u> Float <u>whole Strawberries</u> and serve over ice.
Makes about 3-4 quarts.

STRAWBERRIES in WINE

Amount of Strawberries or Champagne depends on number of people to be served.

Fresh Strawberries - with stems
Champagne (or white wine)
Refrigerate above for several hours.

Using Champagne glasses in which to dip Strawberries, allows each person to heap a small plate with berries from a large bowl. Provide small dishes of sugar for those with a sweeter tooth. Strawberries seem to disappear rapidly as they are dipped in and out of champagne.

The coast of California was a paradise for plant explorers in the early 19th century, resulting in new specimens at the Royal Botanical Gardens at Kew and at the Royal Horticultural Society of London.

8

"...also I found a bery growing lowe at my first landing whiche in bery was muche lik a strabery but of an amber coller, the people eate it for a midsin against the scurvi ... I dried sume of the beryes to get seede whearof I have sent to Robiens of Paris."

from diary of a trip to Russia in 1618

"Hot pescodes, one
began to crye,
Strabery ripe, and
cherryes in the ryse"

John Lydate
15th century

These words mimicked the call of the 15th century London straw-berry street vendor. The words mean hot peas, and a stick of bacon, strawberries and cherries. on the branch. This is the first use of the word Strabery in English writings. (1425)

9

FUNDY DRESSING

2 Tablespoons strawberry jam
1 cup sour cream
¼ cup mayonnaise

Mix together until well blended
Dressing can be thinned with
1 or 2 teaspoons light cream.

Hairpins are used to pin down strawberry
"daughters" (runners). Can be purchased
wholesale in Appleton, Wisconsin.

SOME WILD IDEAS

The wild strawberry still grows
in the meadows, along the dunes,
on hillsides and where the
pasture edges into wooded areas.
Why not bring some plants home
and put around small shrubs
or as an edging for your flower
garden - as beautiful flavourful
ground cover. These plants
make attractive and practical
hanging baskets, because of
their runnering ability and fruiting
on daughter plants.

DEVONSHIRE CREAM
with STRAWBERRIES

Some of my English friends reminded me that <u>THE</u> ambrosial delight is Devonshire Cream - with strawberries. It can be served over the fresh berries or on warm scones covered with strawberry jam.

This recipe may be difficult to make unless you have a cow - or are as lucky as I am to have a dairy herd next door.

A friend from Devon gave me her recipe and I share it with you, although I did not

follow her instructions completely.
She said, "only use cream from
Jersey or Guernsey cows," and
the lovely black and white cows
that now seem a part of Cranberrie Cottage, are Holsteins.
In their defense, their cream
was ultra superior and made
exquisite crusted clotted Devonshire cream.

I let the non-homogenized
milk rise overnight in refrigerator
in a large bowl. In the morning
I skimmed off the cream and
put it in the top of a double
boiler. Over very low heat the
cream steamed for six - that's
right - six hours. It is important
that no water touch the cream.
A thick crust formed on top.

When cream was barely
lukewarm, I put it back in
the refrigerator to chill.

This is an old recipe - try
it at least once and you'll never
forget Devonshire Cream with
Strawberries.

Kingsley Brown of Antigonish, N.S. tells
of a feast of strawberries when he
was a prisoner in a German jail in
Amsterdam in 1942. He said the
berries were huge, "And the
cream was not the stuff we
Canadians call cream. It was the
rich, heavy, yellowish cream
the Europeans call clotted, the
kind known in Britain as
Devonshire cream."

from Legion Magazine, 1982
            Canvet Publications, Ltd.

Prior to WWII, most strawberry
production in the U.S. was along
the East Coast, where labor
was cheap and plentiful. Since
WWII such labor was in short
supply and production centres
moved west - California and
the Pacific North West took over
the commercial market by the
60's. After WWII when the Japanese
were removed from Pacific coast,
the California strawberry industry
almost disappeared

CALIFORNIA PARFAIT

½ cup sugar — 6 egg yolks
¼ cup brandy - ¼ cup Sherry
Pint sliced strawberries
Using a double boiler, combine
egg yolks and sugar and heat
slowly. Add brandy and sherry.
Stir until thick. Remove from
heat, stir in sliced strawberries.
Spoon into parfait glasses.
Serve immediately.

STRAWBERRY FROST

1 cup thawed and drained frozen
            Strawberries
2 mashed bananas
Whipping Cream (optional)

Beat together 2 egg whites and
¼ cup sugar, to peaks. Whip
1 cup chilled evaporated milk
until thick. Fold into egg whites.
Add about 1 teaspoon lemon juice.
Fold in drained strawberries
and mashed bananas. Spoon
into molds or trays. Freeze
quickly. Serve with tiny
ginger cookies.

STRAWBERRIES with SUGAR

Sprinkle bowl of Strawberries
with granulated sugar. Set aside
several hours before serving.
(Top with Sour cream, or
soft ice cream, or chopped
nutmeats. Drizzle an orange
flavoured liqueur over top.)
I like the first berries with sugar
alone.
No domestic plant is more complex
biologically or more sensitive in
its adaptability, than the Strawberry.

BAKED STRAWBERRY CUSTARD

Prepare pastry for unbaked
bottom shell.
Combine ½ cup sugar, ½ tsp. salt
and 1 T. flour. Add 3 beaten
eggs. Scald 2½ cups of milk
and add to mixture.
Spread 1½ cups sliced sweetened
STRAWBERRIES, on top of hot
mixture that has been poured
into shell. Bake about 45mins.
in a preheated 350° oven.
Cool well before serving.

The wild strawberry F. <u>Virginiana</u>
is native in Canada as far North
as the 64th parallel, the latitude of
Iceland, Stockholm and Helsinki.

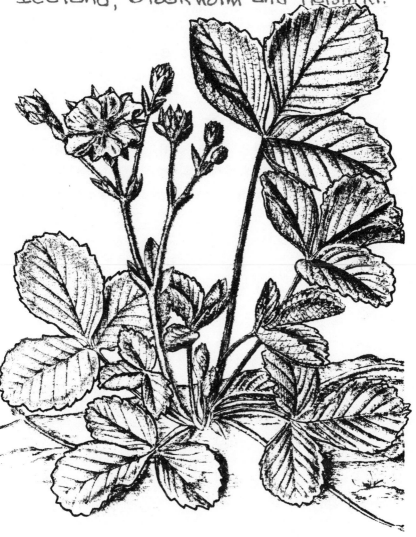

DRAMBUIE
(AN DRAM BUIDHEACH)

Whenever something was expensive, or really satisfied, my Calvinist Scottish ancestors felt compelled to shy away from such perceived decadence. In a lifetime I have not yet unloaded the whole burden of my heritage. But - now and then I do insist on an indulgence - and what could be more indulgent than using Drambuie in a special recipe with strawberries - a custom bedded deep in my collective past.

Drambuie is compounded from a secret formula of Scotch whiskey, heather honey and herbs. Legend is that the recipe was given To a family of MacKinnons, who lived near Edinburgh in Scotland, by Bonnie Prince Charlie, as a thank-you for assisting in his escape to France.

The connection between Scots everywhere, and Drambuie, is an enduring one. Here in the maritime province of Nova Scotia (the name means 'New Scotland') the heather heritage is cherished and perpetuated, as it is by Scots in other parts of Canada, and in the U.S.A., especially in the Carolinas. I wonder how many people, even Scots, know that the word Drambuie is a corruption of the Gaelic — AN DRAM BUIDHEACH, meaning, "The drink that satisfies."

Drambuie is one of the most popular liqueurs in Canada and the U.S.A. It is a culinary delight to mix fresh strawberries and Drambuie — and to revel in such decadence.

A TASTE of CLASS
(Strawberries with Drambuie)

Combine 1¼ cups milk to 1¼ cups cream. Stir in 6 Tablespoons sugar. Heat and stir. When the sugar is dissolved, stir in Drambuie to taste.

Dissolve 3 Tablespoons gelatin in ¼ cup hot water. Slowly add to milk mixture. Crush about ½ cup almonds and fold into the mixture. Pour into lightly oiled mold. Chill until set.

FRUIT PURÉE TOPPING:

Simmer 1 quart strawberries until tender. Sweeten to taste. Purée in blender. Stir in ¼ cup Drambuie. Chill well before serving with the gelatin mold.

In the 16th century, public botanical gardens and garden businesses were popular, and used primarily for the study and sale of strawberry plants. Inadvertently, the early gardens furthered the advancement of agriculture.

20

"When I was last in Holborne, I saw good strawberries in your garden there. I do beseech you to let us have a mess of them. King Richard III (III, iv)

In Scotland and England the "street cry of the Strawberry vendor" was commonplace in the 17th and 18th centuries. It can still be heard today in Ireland.

SYRUP DRAMBUIE

1 cup Sugar
¼ cup Drambuie
1 cup Strawberry juice
¼ cup Lime juice

Combine sugar and strawberry juice. Heat slowly. Stir in lemon juice and Drambuie. Cool well before using.

FRESH FRUIT CUP

1 c. strawberries— ½ fresh Pineapple
3 ripe bananas — 3 oranges
2 T. lemon juice — sugar

Peel and dice pineapple. Cut the bananas into thick slices and separate oranges into segments. Wash and hull strawberries. Mix fruits with lemon juice. Sugar to taste. Chill in refrigerator in covered container until ready to serve.

DRAMBUIE DECADENCE

Wash, stem and halve 4 cups of strawberries. Cover with Confectioner's sugar and drizzle 6 Tablespoons Drambuie over the sugared berries. Chill well.

Beat 2 egg whites. When stiff beat in 4 Tablespoons powdered sugar. Fold into chilled berries.

Serve in sherbet glasses. Garnish with fresh or candied mint sprigs.

22

STRAWBERRY VINEGAR I

There are several versions of this drink that was popular in ancient times, even as today.

Fill a quart jar with ripe strawberries. Add white vinegar, pressing the berries gently. Fill jar with vinegar.

In about 1 month, strain and bottle the liquid. Combine with soda water for a delicious summer drink. To serve, add 2 cups sugar or honey, for each pint of vinejar/juice.

Home gardeners who have read Rachel Carson's book are more likely to pray than to spray.

This recipe should inspire you to have a herb garden - or to use more herbs daily.

MAIBOWLE

This wine, drunk in Germany on the first day of May, dates back hundreds of years.

   1 bottle dry white wine,
   1 cup whole strawberries
   12 sprigs of Woodruff
   1 tsp. white sugar

Put in a jug and leave 1 hour. Strain and serve chilled, with jug surrounded by crushed ice.

The wild strawberry was found in most kitchen gardens. A small plant with berries -cleaned- is a common garnish for the wine.

Three strawberries are shown (numbers 6, 7, and 8) in this illustration from "The Kitchen Garden," in *Paradisi in Sole* by John Parkinson (1629). The title of the herbal is a pun on the author's name: Park-in-the-sun.

24

It was the breeding experiments with the Strawberry that opened the door to scientific inquiry into all plants.

STRAWBERRY-LIME

Soften 1 pint lime sherbet by removing from freezer for several hours.
Cook about 3 cups Strawberries about 15 minutes — or until they have thickened. Spoon the purée into a bowl and chill.

In a 1-quart mold, put a layer of the sherbet. Press it to cover the sides too, and freeze immediately until it is firm.

Combine ½ cup sugar and ⅓ cup water and cook until smooth. Bring to a boil and cool until barely lukewarm. In another bowl beat 3 egg yolks and add the sugar syrup. Put in

STRAWBERRY-LIME - (continued)
top of double boiler and
cook over water for about
20 minutes. Then beat with
electric mixer until cool. Chill
covered about 1 hour.

Put the strawberry purée
into the yolk mixture and add
½ cup cream whipped to
soft peaks. Spoon into the
sherbet mold. Cover with
waxed paper overnight.

Serve with glacéed Straw-
berries.

This dessert is worth the
work involved.

SIMPLE SUMMER SOUP

1 pint strawberries
Sugar to taste (¾ - 1 cup)
¾ cup cream - 1½ pint water

Combine berries with sugar
and crush. Add cold water
and chilled cream. Mix well
or blend. Its ready to serve.

THREE- FRUIT JAM

2 cups Strawberries
1 cup crushed pineapple
2 small oranges
5 cups sugar - ¾ cup Water
1 - 1¾ - oz. pkg. fruit pectin

Crush strawberries. Grate orange peel, crush segments and combine with strawberries. Add pineapple and sugar.

Over medium heat, sprinkle fruit pectin with water. Boil one minute, while stirring. Stir this mixture into fruit. Blend well. Ladle into small jars and cover. Set at room temperature for about 24 hours. Freeze. Keeps one year well.

"In the Garden, strawberries delighteth in the sunny places and with good fertilization yieldeth a great deal of large fruit." (1631)

No-Cook Jam

1 quart crushed strawberries
4 cups sugar
2 T. lemon juice
½ bottle liquid fruit pectin

Thoroughly crush enough strawberries to measure 1 ¾ cups
into a large bowl. Add 2
Tablespoons of Drambuie liqueur.

Mix sugar into fruit and let
stand 10 minutes. Add citrus
juice and pectin. Continue to
stir for 3 minutes. Pour quickly
into jars. Cover immediately
with tight tops. Let the jam
set about 24 hours. Freeze.

Recipe makes about 5 cups.

"Strawberry Fever" (1858-70)
No other fruit ever caused
such a furor in the United
States. The strawberry was
considered to be the "fruit
of fruits."

STRAWBERRY MOUSSE

Purée Strawberries to make 1 cup
½ c. water — ¼ c. Sugar — salt
1 tsp gelatin — 1 pt. whipping cream

Soften gelatin in cold water. Add
sugar and heat slowly. Add this to
the fruit. Pour into bowl. Chill
until thickened. Then beat well
and fold in whipping cream.
Cover and freeze.

The wild strawberry of No. America
surpassed all others in beauty, size
and flavour.

31

STRAWBERRY BUTTER

Spreads can be made with many berries as well as with parsley and other herbs.

Mix together equal amounts of butter and strawberry jam. Beat until smooth. Cover and keep in cool spot.

STRAWBERRY

SIMPLE STRAWBERRY PIE

3 cups Strawberries
1 cup Sugar — 1 T. cornstarch
Pinch of Salt — 1 T. butter

Line pan with bottom crust. Add above ingredients except butter, adding that on top of filling. Cover with top crust. Bake in very hot oven 10 mins. at 450°F. Reduce to 350°F for another 30 mins.

"Straberies, they are as good and as great as those which we have in our English gardens."
Sir Walter Raleigh

32

GRAPEFRUIT COCKTAIL
(with Strawberry)

Cut grapefruit in half, carefully remove pulp, leaving white skin as lining. Place shells in cold water to firm them.

Mix pulp with strawberries, crushed and sweetened to taste. Sprinkle sugar on the top. Chill.

To serve, fill shells with the mixture of fruits, placing one large Strawberry in center.

WINE 'N VINEGAR BERRIES

1 pint prepared Strawberries
1 cup white wine
5 Tablespoons White Vinegar
Confectioner's Sugar

Cover Strawberries for about
10 minutes with a cheap table
wine. Chill.
Then drain off wine and
put berries in a large serving
bowl. Sprinkle the vinegar
over the top. Toss lightly.
Add sugar to taste and toss
again - lightly.
The combination of wine,
vinegar and sugar enhances
the natural sweetness of the
Strawberries. (The wine is
used here mainly to clean
and freshen the fruit. Water
acts in reverse and also
makes berries watery and
mushy.)
This is an ancient feast revived.

STRAWBERRY RICE

1 cup cooked chilled rice – salt
1 cup heavy cream – 1/4 c. sugar
2 cups strawberries

Add sugar to cream and
whip. Add salt to the rice.
Fold whipped cream into
rice and put into serving
dish. Top with fresh sliced
strawberries.

STRAWBERRY SPRITZER

3 (10 ozs) pkgs. thawed frozen
                    strawberries
2 – (24 ozs.) bottles white wine
1 – (28 oz) bottle soda water

Place 2 pkgs. thawed berries
in blender until creamy.
Combine with the wine and
the 3rd pkg. of strawberries.
To serve – add soda water
and stir. Enjoy.

GATEAU SUISIENNE

2 round layers Sponge Cake
3 cups Strawberries, puréed
1/4 cup Sugar - 7 T. raspberry jam
1 quantity Meringue
    (see page 132 for cake recipe)

Pick out 12 firm red berries
and set aside. Purée rest of
strawberries with sugar.
Spread on one sponge layer.
Sprinkle other layer with
some of the fruit juice and
place over berries. Cover
top with jam.
Place cake on an oven-proof
plate. Pipe fingers of Meringue
au Cuite around top edge and
decorate. Brown in hot oven. Dip
berries in hot raspberry jam and
arrange on top.

GRANVILLE PIE

3½ ounce can coconut-flaked
1 quart French vanilla ice cream
2 Tablespoons Strawberry syrup
1½ cups sliced strawberries
½ cups mint flavoured sugar
          Drambuie

Grease a 9-inch pie pan. Line
with aluminum foil. Preheat
oven to 350°F.
Spread coconut on foil and
bake until barely golden,
about 10 minutes. Cool.
Press onto bottom and sides
of pie pan.

Stir ice cream, adding berry
syrup. Spoon into coconut
shell. Freeze until firm.

Combine Strawberries and
sugar and spoon over ice
cream just before serving.
Drizzle Drambuie over the
top.

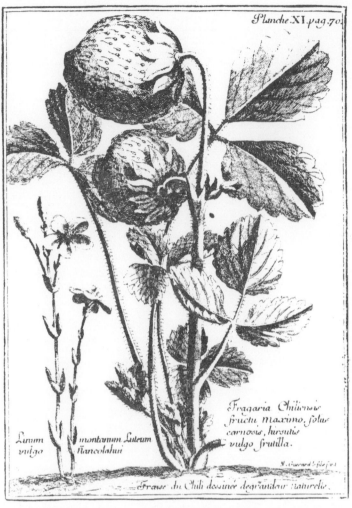

Planche XI.pag.70.

Lunum vulgo *montanum Luteum flancolahui*

Fragaria Chilensis fructu maximo, folus carnosis, hirsutis vulgo frutilla.

N. Guerard le fils fecit

Fraise du Chili dessinée de grandeur naturelle.

    Fraise du Chili or *Frutilla* "being the large strawberry of Chile, drawn after its natural bigness. They plant whole fields with a sort of strawberry rushes [bushes], differing from ours in that the leaves are rounder, thicker, and more downy. The fruit is generally as big as a walnut and sometimes as a hen's egg, of a whitish red, and somewhat less delicious of taste than our Wood strawberries. I have given some plants of them to Monsieur de Jussieu for the King's garden where care will be taken to bring them to bear." From A. Frézier (1717), *A voyage to the South Seas and along the coasts of Chile and Peru in the years 1712, 1713, and 1714.*

Minted Strawberry

1 quart sliced strawberries
2 large bottles soda water
2 bottles white wine
6 sprigs fresh mint
Sugar

Crush the mint in a wooden bowl. Pour wine into large container and add crushed mint. Put in a cool place for 1-2 hours.
Sprinkle sugar over strawberries and add to wine, removing mint before serving.

When serving, put some berries in each glass and about equal amounts of wine and cold soda water.
This can also be served as a punch by combining in one large bowl.
If you have lots of fresh mint, it is a cool garnish.

STRAWBERYE SAUCE

3 cups Strawberries - 2 T. honey
1 cup white wine - 2 cups cream
1/4 cup nutmeg - 1/8 tsp. salt

Mash 2 cups of berries and
slice remaining cup. Heat the
cream but do not boil. Add
wine, honey and nutmeg. Stir a few
minutes, then add mashed berries.
Just before serving add the
sliced strawberries.

This is an interesting recipe
and delicious with chicken or
turkey. Eleven women from
Canada and the U.S.A. sent me
a version of the recipe. Some
used cinnamon but I found
the nutmeg gives a better
flavor. Another variation of
the recipe is in Madeleine Cosman's
book, Fabulous Feasts, published
by Geo. Braziller.
Try it - you'll like it!

STRAWBERRY TEA

The leaves can be used fresh
or dried. For 2 cups, use
about 6 stems with 3 leaves each.
Cover with boiling water and
let stand about 4 minutes.
Remove leaves from the pot
and serve the infusion.
The tea is also refreshing
served cold or added to a
recipe for fruit punch.
Throughout the history of the
strawberry, there is mention
made of the leaves, crowns,
roots and fruit being used to
make teas, syrups, tinctures
and ointments. Even today,
in some areas, strawberry
tea is believed to curb
"bladder infections."
Sometimes the leaves are eaten
with blackberry leaves, fried in
grease, or boiled in water with
fat back added. This is not my
favourite recipe!

STRAWBERRY CRÊPES

Strawberries
2 Tablespoons honey
2 sprigs mint – 2 scented geranium
Whipped Cream   leaves
1 cup sifted flour – 1 tsp. sugar
2 eggs – 1¼ cup milk – salt
2 Tablespoons butter, softened
½ tsp. each grated orange and
                        lemon peel

Soften honey in ½ cup water
over low heat, about 10-12 minutes.
Add strawberries, then mint
and geranium leaves and gently
poach . Cool.
Remove greens. Then gently
remove the strawberries and
carefully mix them into the
whipped cream . Add citrus
grated peel.
Make the crêpes and when
they are ready, spoon the mixture on top. Roll them in sugar
and dribble the minted syrup
over each one .

STRAWBERRY ICING

1 cup sugar   —   1 egg white
1/4 teaspoon cream of tartar
1/2 cup boiling water
2 Tablespoons strawberry gelatin
1 teaspoon vanilla

Combine all except vanilla.
Beat vigorously at high speed.
When peaks are stiff, add the
vanilla.
Recipe covers 8-inch layer cake

STRAWBERRY-RHUBARB JAM

1 quart Strawberries
1½ cups rhubarb, 1" pieces
4 cups sugar — ¾ cup water
1 box fruit pectin

If rhubarb is frozen, thaw.
Crush strawberries, one layer
at a Time, then measure 1½ cups
into large mixing bowl. Finely
chop rhubarb and combine the
fruit.
Mix sugar into fruit and let
stand 10 minutes. Combine fruit
pectin crystals and ¾ cup
water in small pan. Bring to a
boil and boil 1 minute, stirring
constantly. Stir into fruit
mixture and stir about 3 minutes. Pour quickly into jars.
Cover at once with tight lids.
Let stand overnight to set,
then freeze. Jam will keep
2-3 weeks in refrigerator, or
store in freezer.

STRAWBERRY SCANDINAVIAN

Prepare a baked 9-inch crust.
2 cups boiling strawberry juice
1 3-ounce pkg. Danish junket dessert
8-ounce pkg. cream cheese
3 Tablespoons sugar -   Salt
½ teaspoon almond - ¼ cup cream
1 cup sweetened sliced strawberries
Combine junket with the juice
and beat hard for 1 minute.
Cool.
Cream cheese, sugar, salt
and flavouring. Fold in the
cream, whipped stiff.
Save some of the creamed
cheese for garnish and
spread the rest in pie
shell.

Gently fold strawberries
into junket mixture. Spoon
onto top of cheese in pie
pan. Refrigerate at least
4 hours.
To serve - garnish with
the extra creamed cheese.

45

SANTA FE DIP

Mix together in saucepan:
  ½ cup hot coffee – ½ cup sugar
  1 square sweet chocolate
Stir until smooth and creamy,
over medium heat. Do not boil.
Add:
  1 Tablespoon each Kahlúa and water
  2 Tablespoons corn starch
When mixture thickens, remove
from heat.
Add 3 more Tablespoons Kahlúa.
Mix well. Cool.

Clean 3 boxes strawberries,
leaving stems attached. Dry
carefully on paper toweling
for at least 15 minutes before
dipping.

Strawberries can be dipped
and stored between wax paper
or can be individually dipped
as an appetizer or after-
dinner treat.

STRAWBERRY WINE

For over a thousand years wine has been regarded as a food, a beverage, a medicine, a tranquilizer, a tonic and a soporific — even as a religious symbol.

The classic vintners used grapes but we know that centuries ago, wine was made from wild flowers and wild berries

Here are two basic recipes for Strawberry Wine. Strawberry wine is difficult to make as the delicate flavour can so easily be lost by oxidation. It is essential to add 1 or 2 Campden tablets to the wine just before each racking to avoid this problem.

STRAWBERRY WINE I

4 pounds Strawberries
3 pounds Sugar
1 gallon (60 ozs) water
1 teaspoon yeast nutrient
1/4   "   grape tannin
2   "   acid blend
2 Campden Tablets
1/2 teaspoon pectic enzyme
    powder
Wine Yeast
(starting specific gravity
should be 1.090 - 1.095,
acid 60 o/o.)

Crush the ripe firm berries
and add all ingredients
except the wine yeast in
primary fermentor. Add
hot water and stir to
dissolve sugar. Cover with
plastic.
When mixture is cool (70-75°F)
add wine yeast. Stir daily.
Ferment about six days or

until SP is 1.040. Strain.
Siphon into gallon jugs or
carboys and attach the
fermentation locks.
Rack after 5 weeks and
then again in 3 months.
When wine is stable and
clear, it can be sweetened
to taste with sugar
syrup (2 parts sugar to 1 part
water). Bottle, after adding
3 stabilizer tablets to pre-
vent continued fermentation.

Age 1 year before drinking.

"Wine can be considered
with good reason as the
most healthful and the
most hygienic of all
beverages."
                    Louis Pasteur

STRAWBERRY WINE II

3 ¼ lbs. Strawberries
2 ½ " Sugar
1 T. Benerva (3mg. Vit. B tablet)
2 level tsp. Tartaric acid
1 " " grape tannin
1 tsp. ammonium phosphate or
      1 nutrient tablet
Wine yeast - Water, about 1 gal.

Make a yeast starter in a
clean wine bottle with the
juice from a handful of the
strawberries, 2 teaspoons
sugar and 1 cup cold water.
Add yeast and plug bottle
with cotton.

Place about 3¼ pounds of
strawberries in a plastic
bucket and crush with a
wooden masher or hammer.
Add the sugar, the additives,
and 5 pints of water. Stir
well. Add yeast starter.

Ferment 3 days, straining through muslin into a gallon jar. Continue to ferment under an air-lock.

As fermentation diminishes, top up with water. In about 1 month, rack off into another jar, adding 1 Campden tablet. Top with water, and insert a bored cork plugged with cotton or wool or with an air-lock. Maturing should take about 6 months. Wine can be sweetened with sugar, ¼ to ½ pound per gallon.

"Wine is one of the noblest cordials in nature."

John Wesley
(founder of Methodism)

STRAWBERRY HONEY TOPPING

½ cup yogurt – 2 T. honey
Grated rind of 1 lemon
Juice of 1 lemon – 1 cup Strawberries

Chop strawberries and fold into
a mixture of the yogurt, honey
and lemon.

Strawberry blossoms are pure
white and, like violets, in the shape
of little stars. The center is yellow
and this becomes the berry when
the petals fall off.

SUNDAY DESSERT

4 cups small whole strawberries.
Mix with 1 small fresh pineapple,
pieces cut bite-size. (The pineapple should be marinated in
white wine - about ½ cup- for
at least six hours.) Sweeten
with sugar and flavour with
Rose water. Heat 1 cup grape
jelly mixed with wine used to
marinate pineapple. Pour over
chilled fruits.

52

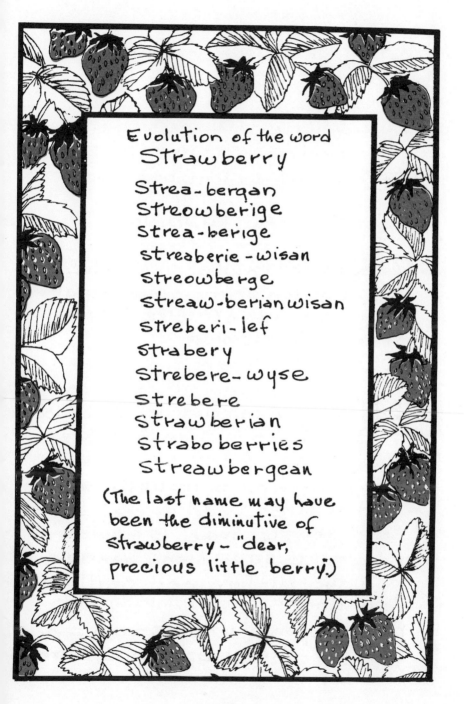

Evolution of the word
Strawberry

Strea-bergan
Streowberige
Strea-berige
streaberie-wisan
streowberge
streaw-berianwisan
streberi-lef
strabery
strebere-wyse
strebere
strawberian
straboberries
streawbergean

(The last name may have
been the diminutive of
strawberry- "dear,
precious little berry.)

Early in the 1300's, monks in western Europe used wild strawberries in their "illuminated" manuscripts. The Madonna was often shown among the strawberries,

CEREAL PUDDING

2 cups Strawberries, halved
2½ cups water — ½ cup sugar
1 cup Cream of Wheat cereal
½ tsp. maple syrup
Whipped cream

Boil strawberries in water for 5-8 minutes. Drain but keep the juice. Add the sugar to the juice and stir until sugar is dissolved. Add cereal and cook gently about 15 minutes. Add maple syrup, and berries, into the mixture and chill for four hours, until firmly set. For company, whip cream and flavour with maple, with berries on top.

without
pollination by
bees, there
would of
course be
no Straw-
berries, or
shortcake
or juices –
or any fruits.

FRESH FRUIT BOWL

2 cups halved Strawberries
5 oranges – 2 unpeeled apples
2 small bananas, thick slices
2 Kiwi fruit - peeled and sliced
1 T. lemon juice – Honey dressing
(see page 52)

Remove orange segments. Add
together all the fruit, first
sprinkling lemon juice on banana
slices. This tastes best if
made several hours ahead
and well chilled. Sweeten with
honey dressing or sprinkle with
vanilla sugar.

55

Early pictures of the Virgin
often showed the symbolic
strawberry. One, in 1400, shows
St. Joseph holding out a strawberry to the Child Jesus; in
another painting, the angels
are gathering strawberries for
the Christ Child on His mother's
knee.

FROSTED STRAWBERRIES

Plan ahead as berries should
be prepared about five hours
before serving. Find a cool spot
but _not_ in the refrigerator!

Select firm perfect berries.
Dip each one in a cup of
white wine, mainly to wash
them. Arrange on absorbent
towelling.
Beat two egg whites about
one-half minute with fork.
Dip each berry in egg white and
then in a cup of vanilla sugar.
Harden on wax paper. Cool.

56

"Some grow on the mountains and woods, and are wild, but some are cultivated and are so odorous that nothing can be more so."   Estinne de Re Hortensi, 1585

STRAWBERRY

STRAWBERRY-LEMON SAUCE

2 cups fresh strawberries
1 lemon -   Sugar to taste

Combine berries, juice and grated rind of 1 lemon in a food processor. Purée until smooth. Add sugar - very small amount at a time and check for sweetness desired. Serve over puddings or ice cream. Cover and keep chilled before and after serving.

The three major Strawberry production areas in Canada: the Maritime Provinces, Ontario and Quebec, and British Columbia.

GINGERED FRUIT SALAD

2 cups orange juice - 1 lime
1/4 cup slivered crystallized ginger
2 cups fresh halved strawberries
1 small Honeydew melon
Sugar

Pour orange juice into a large
container with tight-fitting lid.
Stir in juice and finely grated
lime peel and slivered ginger.
Add halved, or sliced, strawberries to juice mixture, with
wedges of melon. Gently
fold fruit into mixture until
well covered by juice. Taste.
Sweeten if you wish. Seal
tightly and refrigerate until
ready to serve.

STRAWBERRY DAIQUIRI

Blend together, serve over crushed ice:

1½ ozs. light rum
1 scant teaspoon sugar
2 ozs thawed strawberries
and their syrup.
Serve in a cocktail glass

STRAWBERRY-YOGURT DIP

2 cups plain yogurt – 1 tsp. sugar
1 T. finely chopped fresh mint
Peel of 1 lemon, finely grated
Strawberries, whole

Combine yogurt, mint, lemon and sugar. Stir well. Chill until ready to serve. Serve with whole stemmed strawberries for dipping.

Fruit is hidden by the foliage, deep red when mature, consistently large, the size of a plum, fleshy, and of excellent flavor and fragrance.
Basilius Besler (1613)

strawberries were cultivated by masters skilled in the art of gardening.

STRAWBERRY BREAD PUDDING

1 quart milk — 2 T. melted butter
2 c. fine dry bread crumbs
1 c. light corn syrup – 1 t. vanilla
1 c. Strawberry sauce or Jelly
6 T. sugar – 3 eggs separated

Pour milk over crumbs. Blend melted butter with syrup and add well beaten egg yolks. Add vanilla and combine with crumb mixture.
Turn into buttered baking dish and bake at 325° until firm. Spread with generous layer of Strawberry sauce. Cover with meringue: —
Beat egg whites until frothy. Add sugar gradually, beating until soft peaked. Bake until browned. Cool and serve.

Frozen strawberries
progressively lose Vitamin C
— about 50 % - 4 months
— about 70 % - 6 months

The more sunshine, the more
Vitamin C (ascorbic acid).

PORT ROYAL SALAD

2 envelopes gelatin
1½ cups hot water
⅓ cup grapefruit juice
1 cup diced, drained grapefruit
sections
1 cup Strawberries, quartered
Mayonnaise - Crisp Lettuce
6 whole Strawberries

Dissolve gelatin. Add grapefruit juice. Chill. When
slightly thickened, fold in the
fruit. Pour into ring mold
and chill until firm. Un-
mold on lettuce. Garnish
with one of the light
dressings or mayonnaise.
Top with whole strawberries.

SOUFFLÉ AUX FRAISES

1 cup Strawberries, drained
1 Tablespoon lemon juice
1/8 teaspoon salt — Sugar
3 egg whites, stiffly beaten

Crush Strawberries through a
sieve. Add juice and salt and
sugar to taste. Place over low
heat.
Fold stiffly beaten egg whites
into hot strawberry mixture.
Pour into greased baking dish
or individual molds, filling
3/4 full. Set in pan of hot
water. Bake at 375° F for
20 minutes for a soft soufflé.
If you desire a firmer one,
bake 40 minutes at 325°F.
Serve immediately with
cream or sauce.

Strawberry seed was such a rarity
in 1766 that the King of England
purchased seed from Turin — a
pinch of the seed cost one guinea.

62

STRAWBERRY COBBLER

1 cup presifted flour
½ cup melted butter
1¼ cups light brown sugar
salt -   ½ cup milk
             1 tsp. baking powder
             2 cups Strawberries

Grease a 2-quart casserole and pre-heat oven to 350°F. Pour butter into baking dish. Sift together flour, salt, baking powder and ¾ cup of sugar. Add milk. Beat until smooth. Pour into casserole.

Toss rest of sugar with the Strawberries. Sprinkle over the batter. Bake about 30-40 minutes.

63

The CREES called the Strawberry otei-meena; the OJIBWAY called it oda-e-min. Whatever the tribe or the word, the meaning was the same — HEART BERRY.

PARRSBORO PUNCH

Boil 2 cups water and 1 cup of sugar about 10 mins. to make syrup.

Add: 1 cup tea
     2 cups Strawberry syrup
     Juice of 5 lemons/5 oranges
     1 can crushed Pineapple

In about a half hour, strain. Add ice water to make 1½ gals. liquid.

Add 1 quart soda water and 1 cup Maraschino cherries.

Serve in large punch bowl. Serves 50, once!

Rum can be added to this recipe with glowing results.

The Strawberry belongs to the Rose family - Order: Rosaceae; Genus: Fragaria.

Rose Water (1550)

Some do put rose water in a glass and they put roses with their dew thereto and they make it to boile in water, then they set it in the sune tyll it be readde and this water is beste.

Also drye roses put to the nose to smell do comforte the braine and the harte and Quencheth sprites.

Sugar flavoured with roses goes well with strawberries. Best way is to pound white sugar with double its weight of rose petals.

Rosewater Flavouring

The famous Canadian, Madame Jehane Benoit, says that she always adds a "tiny Victorian silver and cut-glass container of rosewater" and a few drops are "enough to bring all kinds of beautiful romantic feelings ... and with strawberries the marriage is perfect."

Recipe for Rosewater

The red wild or Moss rose petals are the best to use. Gather one peck of fresh blossoms for every quart of water. Add petals to cold water and distill slowly over a low heat. Cool and reheat. Fill bottles with the rosewater and cork. After three days, tighten the cork and store at least three months before using.

Sir Hugh Plat (1632) discovered
how to prepare strawberry
jellies and conserves:

" . . . to make gelly of straw-berries, grind them in an alabaster mortar with foure ounces of sugar and a quarter of a pint of faire water and as much rose-water and so boile it in a posnet with a little piece of isinglasse and so let it run through a fine cloth into your boxes, and so you may keep it all the yeare." To make conserve of strawberries: " . . . first seethe them in water, and then cast away the water, and straine them; then boile them in white wine . . . to a stiffness, ever stirring them up and downe, and when they be almost sufficiently boiled put in a convenient proportion of sugar; stir all well together and after, put it in your galley pots . . . "; or a second way: " . . . straine them being ripe, then boile them in wine and sugar till they be stiff." Plat did not specify the quantity of strawberries required in the recipes.

Florida Pie

Prepare 1 pint Strawberries and fill a cooked pie shell.

Combine ½ c. water with 1½ c. sugar and ½ tsp. cream of tartar in a saucepan. Cook gently about 20 minutes. Drizzle over the berries.

Beat whites of 3 eggs, adding ⅛ tsp. salt. When peaks are stiff add 1 tsp. almond extract. Fold gently over top of pie filling. Cool well.

Garnish with finely sliced Strawberries and slivered almonds.

Hybridization means interbreeding, crossing of species, varieties, etc.

① Crossing of two species
② Crossing of two descendants
③ Back crossing to original parents and reselection

Result = a new cultivar, or variety of strawberry

O Lord, how manifold are thy works! in wisdom hast thou made them all: the earth is full of thy riches

PSAL. 104 24

STRAWBERRY SHAKE
(for weight watchers)

3/4 cup water - 1/3 cup dry milk
1/4 tsp. coconut extract
1/8 tsp. almond extract
1 tsp. vanilla - artifical sweetener
1 cup frozen strawberries
3 ice cubes

When all but last two items
are blended, add strawberries
and ice cubes.
Variation: use 1/2 cup evaporated
skimmed milk and 1/4 cup water.

Captain Vancouver of H.M.S. Discovery
sailed along the coast of California
in 1792-3-4. The officers were
served the "berries" found all
along the shore. Aboard the ship
was the famous Scottish botanist
and surgeon, Archibald Menzies of
Styr. He had a greenhouse on
the quarter-deck and collected
botanical specimens from all over
the world. The "berries along the
shore" were strawberries.

UPSIDE DOWN PUDDING

12 crisp crackers, crushed
1 cup milk — 1 T. butter
½ cup sugar — 3 egg yolks
3 egg whites — 3 T. Strawberry
                                   Preserves
Juice and grated rind of 1 lemon

Soak cracker crumbs in milk.
Cream butter and sugar, add
lemon juice and rind. Stir in
beaten egg yolks. Combine
with crackers. Fold in egg
whites stiffly beaten.
Butter a deep pudding dish
and put Strawberry jam on
the bottom. Pour mixture
over jam or preserves, and
bake 30 minutes at 350°F.
When cooled, turn pudding up-
side down and serve cold with
Strawberry ice cream that
is slightly soft.

First refrigerated shipment of Strawberries was in Cincinnati, Ohio -
4000 qts. a day — 1843.

71

The strawberry has been produced commercially in China
for less than 100 years. Seven
species grow wild across China.

Fruits with peel are preferred in
China, for sanitary reasons.

STRAWBERRIES JUBILEE
2 cups Strawberries
2 T. mint flavoured sugar
1 pint French vanilla ice cream
1 can black sweet cherries
3/4 cup black currant jelly
1/4 cup brandy

Mix strawberries and sugar.
While these chill, put small
balls of ice cream on foil
and place in the freezer.
Now crush the berries, drain
the pitted cherries and add.
Melt the currant jelly and
add to fruit, simmering about
5 minutes.
Carefully pour brandy over

STRAWBERRIES JUBILEE (cont'd)

the fruit but do not stir or
the brandy will not flame.
Just before serving, put ice-
cream balls in a bowl.
From a chafing dish, pour
fruit over the ice cream, first
igniting the brandy, making
a festive flaming strawberry
Jubilee.

1 pound fresh Strawberries
= 160 calories
  3.5 grams Protein
  2.2 " fat
  26.5 " CHO
 water content - 89%
1 cup = 149 grams = 54 calories
 Vitamin C = 89 milligrams

COLD STRAWBERRY PURÉE

When the weather is hot, cold fruit purées are often preferred to hot soups. They should be served in cups with a few pieces of fruit on top.

Boil ½ cup granulated tapioca in 6 cups of water and ½ cup raspberry juice. When mixture is transparent, add 2 cups sliced strawberries and sugar to taste. Cool thoroughly.

Serve ice-cold in glass mugs or punch goblets for a sip of summer soup.

70 ogo of the U.S. production of
strawberries come from California.
(70 ojo - fresh ; 30 ojo - processed)
Coastal regions produce 95 ojo of
the total.
   California averaged 19 tons per
acre for past ten years. Michigan
averaged 2.7 tons and Florida
had 6.4 tons per acre.

NATIVE CALIFORNIAN.

STRAWBERRY FOOL

2 pkgs. (10 ozs. each) frozen Strawberries
1 cup dry almond macaroons,
                        crumbled
2 cups heavy cream, whipped

1) Thaw strawberries, following
directions on label. Purée in
blender on low speed, just
until smooth. (or press through
a sieve). Turn into serving
bowl and refrigerate.
2) Just before serving, fold
macaroons into whipped cream;
spoon on top of strawberry
purée. Fold gently together,
leaving streaks of red and
white.

Who were the Strawberry Wives?
"Two or three great straw berries at
the mouth of their pot, and all the
rest with little ones."
                Francis Bacon (1624)

76

It is said that country people used to follow the cows to see which berries or plants they avoided. Don't do this. With apology to my Holstein friends next door, it has also been said that "cows are notoriously stupid." On the other hand, we know they do love the wild strawberries, so they must have some smarts!

KIKI

STRAWBERRY SLUSH

1 cup light Rum
1 can (small) frozen lemonade
1½ cups STRAWBERRIES
2 Tablespoons Confectioner's Sugar

Combine Rum and lemonade and blend quickly. Add Sugar and Strawberries. Blend until slushy.

MOUSSE GLACÉE aux FRAISES

1 quart Strawberries — Juice of ½ lemon
2 cups heavy whipped cream
2 cups Confectioner's Sugar — sifted

Force berries through sieve. Stir in
sugar and lemon juice. When sugar
is dissolved, fold in whipping
cream. Fill mold with the mixture.
Place a layer of wax paper over
the top and fit cover on tightly.
Do not open for at least 3 hours.
If you have a deep freeze, place
the mold in the coldest part for
the next 2 hours..
Color can be heightened by adding raspberry jelly or juice.

Geo. Washington compared the gardener to the general: "how much
more delightful to an undebauched
mind is the task of making improve-
ments on the earth, than all the
vain glory which can be acquired
from ravaging it."

78

IS FOR

LAZY SUMMER PIE
2½ cups Strawberries
1 quart Strawberry ice cream
2 egg whites — ¼ cup Sugar
¼ tsp. Cream of Tartar
Graham Cracker Crust

Prepare and bake crust. Cool.
Fill with about 1¾ cup of
berries. Cover with softened
ice cream. Slice rest of the
berries on top.
Combine egg whites, sugar
and cream of tartar. Whip
well. Cover filling and brown
quickly on centre rack of
oven.

STRAWBERRY VINEGAR II

Combine about 3 quarts cleaned and hulled strawberries with 2 cups cider vinegar. Heat but do not boil. Cool in refrigerator for about 1 week. Strain well.

For every cup of berry juice add ½ cup of sugar. Boil for 5 minutes until clear. Pour into prepared jars. When cool, place in refrigerator. Serve by pouring ¼ cup of juice over ice cubes in a glass. Top with water.

Curly-locks, Curly-locks,
wilt thou be mine?...
(you will) sit on a cushion
And sew a fine seam,
And feed upon strawberries,
Sugar and Cream.   Anon.

"How many strawberries
         grow in the sea?
As many as red herring,
         grow in the wood"

         Dr. Wm Butler (1535-1618)

Strawberry Custard Pie

1 cup Strawberry Jam or Preserves
2 eggs – ½ cup Sugar – salt
1¼ cup milk – 2 T. cornstarch

Make a custard of all except
Jam and egg whites. When
cooked, add jam and pour
into unbaked pastry shell.
Whip egg whites with 2T.
sugar, top the pie. Bake in a
slow oven until browned.
(a pinch or two of cream of
tartar added while beating
will fluff it up)

81

STRAWBERRY hors d'oeuvres

The amounts depend on how
many fritters you need — how
many are coming to the party.

Firm ripe strawberries, with stems
Currant jelly
Finely chopped nutmeats
Eggs, lightly beaten
Very fine bread crumbs

Prepare strawberries and dry
them well. Heat the jelly so
consistency is right for a dip.
Dip each berry in the jelly,
then into a small deep dish
of nutmeats, then into beaten
eggs, and then into bread
crumbs.
Cover a platter with wax
paper, using one that will
fit into refrigerator. Lay each
strawberry on the platter.
Chill for at least 4 hours.
(continued)

Just before serving time, pour salad oil into a deep-frying pan. When it is hot, about 350°-360°F, take each berry by the stem, or use a large holed spoon, and fry berries until they are golden color.

Place each strawberry on a paper towel, then to the serving dish

STRAWBERRY RATAFIA

Strawberries steeped in brandy
make a very special gift.

1 quart strawberries - Rosemary
6 lemon slices – 2¼ cups Brandy

Combine and steep for 1 month.
Keep in a warm daytime spot.
Strain and bottle. Now it is
time to drink the Ratafia.

ORANGE or LEMON SUGAR

Mix ½ cup sugar with ¼ cup of
citrus peel. Store in an air-tight
tin. Sprinkle lightly on fresh
fruit – especially strawberries,
or use as a topping only.
(citrus peel should be finely
grated and dried for one day in an
open dish, before mixing with the
sugar.)

ICED STRAWBERRY SOUP

1 quart Strawberries - ½ c. water
½ c. dry white wine - ½ c. sugar
⅛ tsp. each allspice and nutmeg
1 cup buttermilk.

Prepare strawberries. Set
aside about 8 berries for
garnish. Purée remaining
strawberries in blender
with wine and water. Pour
Purée into large serving bowl.
Stir in sugar, spices and
buttermilk. Chill 4-6 hours.
Just before serving, slice
8 strawberries and stir
into soup. Should serve
4-6 people.

ROYAL STRAWBERRY JAM

Queen Victoria, in her later years, was said to complain that the strawberries didn't taste as good as they had when she was a young girl.
Here is a strawberry jam from Chicago — guaranteed to please royal taste buds whatever the age.

2 quarts firm strawberries
6 cups vanilla sugar
⅛ teaspoon salt
2 T. light corn syrup
¼ cup fresh lemon juice
2 T. Grand Marnier

Combine berries, salt, syrup and sugar in large non-aluminum pot. Mix, then stand from 5 to 6 hours. Do not chill.
Prepare jars and lids.
Heat strawberry mixture slowly until sugar dissolves.
Once dissolved, do not stir.

ROYAL STRAWBERRY JAM (con't)
Boil about 25 minutes until the
liquid begins to thicken. Remove
from heat. Skim foam. Cool.
Stir in lemon juice and liqueur.

Ladle berries into jars, not
quite to the top. (Process as
per manufacturer's directions.)
Cool and store in cool dark
place. Makes between 5 and 6
cups.

Fruit of the woods strawberry cultivated in France during the four-
teenth and fifteenth centuries. The figure was taken from the margin of a
fifteenth-century illuminated painting of medieval Paris. The original is in
the Bibliothèque de la Ville de Paris and was reproduced by Le Roux de Lincy
in his *Paris et ses Historiens aux XIV° et XV° Siècles* (1867).

STRAWBERRY ACADIAN

Clean and hull about 1 cup of fresh strawberries. Cut them into quarters and combine with 1 egg, 1 cup milk, 2 tsp. honey and 6-8 ice cubes. Blend at medium speed and serve immediately.

Variation: Add ½ cup light Rum.

STRAWBERRY FRUITTI

Into a stone jar put one cup of your best brandy, one cup sugar and one cup strawberries. Gently stir until well mixed.

As each fruit comes into season, add it with one cup of sugar, to the crock — one cup for each cup of fruit. Do not add more Brandy. STIR at each fruit addition. Cut up large fruit and pit the cherries.

This is an elegant preserve

89

SAN FRANCISCO GOURMET

1 pint sliced strawberries
3 cups cut rhubarb
Sugar to taste
½ cup water - 1 teaspoon salt
2 cups presifted flour
3 teaspoons baking powder
⅓ cup safflower oil
¾ cup milk
Butter - Nutmeg - Whipped cream

Preheat oven to 450° F.
Mix fruit with 2 cups sugar
to taste and water, in a
square baking pan. Cook
for 5 minutes. ——
Combine flour, baking powder,
salt and 2 tablespoons sugar.
Stir in oil and milk until
mixture is moistened. ——
Drop batter by spoon onto
hot fruits to make 9 biscuits.
Make dent in each one for
butter, nutmeg and sugar. —
(continued)

Bake about 25-30 minutes.
Serve while warm, topping
with whipped cream — or
plain sweet cream.

Note: Frozen fruits can be
substituted. Reduce sugar
to ½ cup and omit ½ cup of
water.

STRAWBERRY-RHUBARB JELLY
(No-cook method)

1½ quarts strawberries
1 pound rhubarb, fresh or frozen
6 cups sugar — ¾ cup water
1 box fruit pectin crystals

Thoroughly crush the strawberries,
one layer at a time.
Place crushed fruit in jelly
bag and squeeze out juice.
Measure 2 cups. Finely chop
rhubarb. Repeat as above and
measure 1 cup.

Combine juices in large bowl.
Add sugar. Mix well; let stand
10 minutes. Combine water and
pectin crystals in small pan.
Bring to boil, stirring constantly.
Stir into juices — stir longer, about
3 minutes. Pour quickly into
jars, let stand overnight and
store in freezer - or in refrigerator for 2 to 3 weeks.

PRESIDO COOLER

Simmer about 6 pounds small strawberries in 1 cup water. (You may need more water but don't let berries get watery.) Purée. Add 2 cups sugar to each 3 cups juice.

Combine in cheesecloth ball:

  10 cloves   1 blade Mace
   1 cinnamon stick - about 4"
   Peel of 1 lemon and 1 orange

Add to juice, bring to boil. Simmer about 30 minutes. Remove from heat, and remove spice bag. Bottle or freeze.

Dilute drink with ice water to serve.

Variation: Add 2 cups rum and serve over crushed ice.

This is a lovely cooling drink It brings back wonderful memories of San Francisco in the 50's.

93

OHIO OMELET with STRAWBERRIES

3 eggs - 1 Tablespoon light cream
salt   - 2 Tablespoon butter
           1/4 cup sour cream
           1/2 cup strawberries
           Confectioner's sugar

Beat eggs well, then add the cream and a dash of salt. Beat again, about 30 seconds.

Using a 10-inch skillet, melt the butter, then pour in the eggs. Stir once or twice. Keep omelet from sticking by shaking skillet back and forth.

While eggs are still soft, add 1/8 cup sour cream and 1/4 cup strawberries. Fold over. Remove to serving dish.

Top with rest of strawberries and sour cream. Sprinkle with sugar.

STRAWBERRY CREAM PIE

9" graham cracker shell
2 cups Strawberries - 3 T. honey
⅓ cup milk — 1 envelope gelatin
1 cup cottage cheese
1½ cup Yogurt

Combine milk with gelatin,
dissolved in a bit of water. Warm
and stir. Combine
cottage cheese,
yogurt and honey
with gelatin mixture
in a blender. Whirl
quickly. Chill ½ hour.
Put strawberries,
sliced, in graham
cracker crust. Add
mixture. Chill about
8 hours.

Try a variety
of crusts.

FAST STRAWBERRY FRITTERS

Dip each strawberry in Buttermilk pancake mix, then deep
fry. When still warm, dust
with icing sugar and serve.

FREDERICTON FRITTERS

1 quart strawberries — 1 egg
1 cup milk   — ¼ cup sugar
1 T. butter   — 1½ cup flour
3 tsp. baking powder - almond, salt
Confectioner's sugar or Whipped
cream

Pat washed berries on paper
toweling. Chill. Beat together
milk and egg, add sugar, butter
and almond extract. Stir into
mixture, sifted flour, baking
powder and salt. Stir well.
With fondue forks, spear berries
dip in mixture, fry 6-8 at a time
in deep hot fat (about 375°) for
about 30 seconds. Serve,
using whipped cream as a
dip, or dust with sugar and
cool.

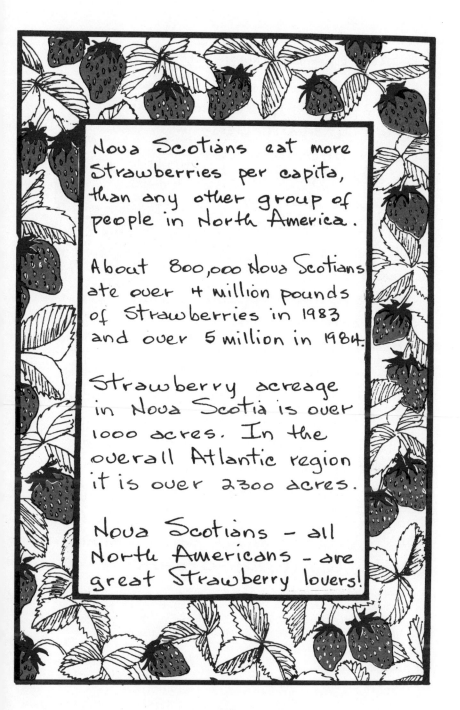

Nova Scotians eat more Strawberries per capita, than any other group of people in North America.

About 800,000 Nova Scotians ate over 4 million pounds of Strawberries in 1983 and over 5 million in 1984.

Strawberry acreage in Nova Scotia is over 1000 acres. In the overall Atlantic region it is over 2300 acres.

Nova Scotians – all North Americans – are great Strawberry lovers!

CORNELL CRUNCH

½ cup presifted flour -
1 cup Instant oats -
1 cup brown sugar-
½ cup butter -
1-pound can of
Strawberry filling-
Whipped cream-

Grease a square pan. Pre-
heat oven to 350°F.
Combine sugar, flour and oats.
Cut in butter. When crumbly,
press half of mixture into pan.
Cover with strawberry pie
filling. Sprinkle remainder
of crumbs over the top.
Bake for 45 minutes. Serve
as squares on individual
plates, topping each square
with whipped cream.

*"I have often been astonished at our indifference to the memory of those
preceding us who have introduced useful (plants) into our country, the fruits
(and beauty) of which we enjoy today. The names of these benefactors are
chiefly unknown, yet their benefits continue from generation to generation
..."*

Jacques H.B. Saint-Pierre
Etudes de la Nature XIII (1784)

FROZEN SALAD

Mash strawberries very fine.
Add half as much sugar as fruit.
Let stand until syrup forms.
Freeze in crank freezer or in
tray of refrigerator. Stir, while
berries are freezing. Serve in
slices on bed of lettuce.
Sprinkle with flavoured
sugar.

BERRY MUFFINS

1 cup Strawberries - quartered
2 cups flour — ¼ cup Sugar
4 tsp. baking powder — 2 T. butter
1 cup milk — Sugar for top

Sift dry ingredients together.
Blend in butter with fingers - do
it quickly. Add milk and berries.

Bake about 15 minutes in hot oven.

"Strawberries be much eaten at
all men's tables in summer for
the pleasantness of them."

STRAWBERRY CUPCAKES

1 cup Strawberry jam
Cream together ½ cup shortening
with 1 cup sugar. Add 2 beaten
eggs. Mix well. Sift 2 cups
cake flour, ½ tsp. salt and
1 tsp. each nutmeg and cinnamon.
Fold mixtures into Strawberry
jam. Bake in greased, floured
pans about 20 minutes at 370°F.

In a sense the strawberry is a New
World fruit as present species are
mainly native to the Americas. The
strawberry grew wild for centuries but
the history of the cultivated present-
day berry, dates back scarcely more
than 2 centuries in French and English gardens.

SUMMER SQUARES

1 cup sifted flour — ½ cup nuts
¼ cup light brown sugar
½ cup melted butter.
Stir together after chopping the
nutmeats. Bake at 350°F for
about 20 minutes.

Now combine:
  2 egg whites— 1 cup sugar
  2 cups strawberries - 1 cup cream
  2 Tablespoons lemon juice
Blend at high speed for 10 minutes. Fold in whipped cream.

Sprinkle most of crumbs in the
pan. Spoon mixture over the Top.
Then put crumbs over mixture.
Freeze overnight. Garnish with
sliced strawberries when
served.

A SIP OF SUMMER

This wine and strawberry experience happened to me back in 1969 - at a reception in Utica, New York. Since then I've served it many times.

Prepare strawberries. Chill them whole in white wine overnight. When ready to serve, place the wine-marinated strawberries in a large punch bowl. Cover berries with white wine — or champagne. Use a large block of ice to properly cool.

Garnish with a sprig of mint or Rosemary.

Bulletin: Strawberries available fresh from Florida to Alaska several months of year; In some parts of West coast, strawberries available fresh every month of the year.

STRAWBERRY FOG

1 jigger gin — ½ tsp. sugar
4 crushed Strawberries
Juice of ½ large lemon

Shake well with ice.
Strain into glass. Fill up
with soda water.

In Great Britain the strawberry
is a symbolic ornament, indicat-
ing rank; example, the golden
strawberry leaves on coronets of
dukes, marquises and earls.

ST. CROIX SHRUB

Combine 2 qts. strained Strawberry juice with 4 qts rum
and 1 lb. loaf sugar. Stir
until sugar dissolved. Cover
and let stand for several days
Strain, bottle and cork tightly.
Optional: ½ pint lemon juice added.
Can be served hot or cold with
plenty of crushed ice and fruit
garnish.

HALIFAX PUNCH

1 quart Strawberry Sherbet
1¼ cup Sugar — 1¼ cup lemon juice
1 cup Orange juice
4 cup chilled Cranberry cocktail
2 - 28 oz. chilled Gingerale
Whole Strawberries

Soften sherbet and spoon into pitcher. Add citrus juices. Let stand until sherbet melts. Add sugar - mix well - chill.
Freeze ice in a ring mold. Place in punch bowl and add sherbet mixture, then cranberry and gingerale. Float whole strawberries - garnish with sprigs of fresh chilled mint.

105

SUNDAY STRAWBERRY

12 Ladyfingers - Soft Custard
¾ cup strawberry jam
½ cup sweet white wine
1 cup heavy cream
¼ cup toasted slivered almonds

Split Ladyfingers and put back
together with jam. Arrange
one layer in shallow dessert
dish. Pour wine slowly over
Ladyfingers and chill 1 hour.
Pour soft custard over the
layer and top it with stiffly
peaked whipped cream.
Scatter almonds over top and
chill. If in season, top with
small whole strawberries, or
drizzle strawberry syrup
over almonds.

"Humblest born, yet earliest, most beautiful
and welcome in their season. Everybody ought to eat them; they are
available and abundant..." (1889)

Early American colonists wrote re
strawberries "as many as would
fill a good ship."

STRAWBERRY JELLY ROLL

Use sponge cake recipe on page 132 . Grease a large shallow roasting pan and line with wax paper. Spread batter evenly about 1/4" thick onto the wax paper. Bake 8-10 minutes in 450°F oven.

Sprinkle top of a wooden board with powdered sugar. Turn cake out on table and cool with the pan inverted over it. Remove pan and paper and trim cake. Spread generously with strawberry jam. Roll up. Place on a dessert platter, edge side down. Sprinkle with Confectioner's sugar.

As early as 1000 AD. and in 1484, strawberries were mentioned in books about the medicinal virtues of plants.

STRAWBERRY MINT PUNCH

Combine 1 quart water and 2 cups sugar. Boil about 20 minutes. Wash 12 sprigs mint. Add the washed mint, cover. Leave about 10 minutes. Strain.

Combine above with 1 cup strawberry juice, 1 cup orange juice and juice of 8 lemons. Cool thoroughly.

Pour into punch bowl, add 1 pint (about 2 cups) Claret wine. Chill with large piece of ice. Garnish with fresh mint and whole strawberries. This punch can be diluted after adding juices, with 1½ cups boiling water.

The Canadian Scarlet strawberries were often referred to as Americana.

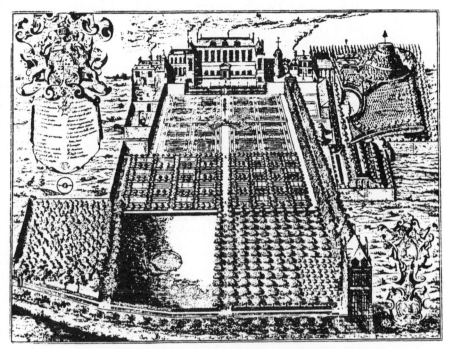

Jardin du Roy pour la Culture des Plantes Médecinales in Paris, as it appeared under directorship of Guy de la Brosse (1636). Here, the *Fragaria americana* (strawberry of Canada), carefully cultivated early in the seventeenth century, became one parent of the garden strawberry and, one hundred years later, the *Fragaria chiloensis* (Chilean strawberry) became the other. Thus the garden figures strongly in the history of the strawberry, and continues to this day as a center of applied botanical research.

Wife, into thy garden, and set me a plot
With strawberry roots, of the best to be got:
Such growing abroade, among thorns in the wood
Well chosen and picked, prove excellent good.

Five Hundred Pointes of Good Husbandrie (1580)
Thomas Tusser

Strawberries were used to landscape the garden of Henry VIII at Hampton Court.

Historic Gardens

From 1710-1750, ANNAPOLIS ROYAL, N.S. was the seat of an English Governor. In the archives are references to a governor's garden. Such a typical garden of the period was reproduced in the recently completed 10-acre Historic Gardens in ANNAPOLIS ROYAL.

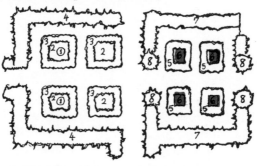

1. Yellow Transparent Apple
2. Day Lilies
3. Boxwood
4. Hill's Yew
5. Mixed Herbs
6. Wild Strawberries ✓
7. Upright Japanese Yew
8. Unicorn Arborvitae

For centuries strawberries were planted as groundcovers in beds of flowering shrubs. In 1368, 12,000 plants were purchased for the Royal Gardens of the Louvre, and wild strawberry plants were cultivated in the Gardens of Versailles.

The Barbery, Respis, and Gooseberry too
Looke now to be planted as other things doo.
The Gooseberry, Respis, and Roses, al three
With strawberries under them trimly agree.

In 1955, 8 Nova Scotia strawberry nurseries planted virus-free stock. The next year they were sold - the first such plants available in Canada. Another Strawberry first!

STRAWBERRY JUICE

Wash and hull berries. Place a layer in bottom of kettle and crush. Add a second layer and crush. Continue until berries are all crushed. Heat about 5 minutes, until berries are soft. Do not boil.
Strain. Then strain again, this time using a jelly bag. Fill prepared jars. Process about 20 minutes in a hot water bath at 188°F.

CROP YIELD
1984
P.E.I. 1,600,000. qts.
N.B. 2,000,000. "
N.S. 5,000,000. "

PARKER'S COVE PUDDING

½ pound marsh mallows
2 cups milk, chilled
1½ cups crushed strawberries
2 Tablespoons lemon juice

Heat marshmallows with two Tablespoons milk over low heat. Fold and fold until marshmallows are almost melted. Remove from heat and continue folding until mixture is fluffy and creamy. Cool to luke warm, then blend with the remaining milk, the crushed strawberries and the lemon juice. Pour into freezing tray and freeze at the lowest temperature, stirring all the time. When firm, beat vigorously and continue to freeze.

"American strawberries are as delicious as any in the world... so plentiful that very few people take the care to transplant them."

History and State of Virginia·1765

STRAWBERRY- PEACH JAM

1¾ cup prepared fruit
   (1 pint strawberries
    ¾ pound ripe peaches)
4 cups sugar — 2 T. lemon juice
½ bottle liquid fruit pectin

Follow directions for Strawberry
Rhubarb jam on page 44.
(Peel, pit and chop peaches
very fine)

GOOSEBERRY-MELON
(with Strawberries)

1 cantaloupe melon
1 cup ripe gooseberries
3 bananas – 2 cups Strawberries
Juice of 1 lemon – 1/4 cup sugar
1/4 - 1/3 cup Cointreau
Mint sprigs

Remove seeds from melon. Scoop out flesh with small scoop. Combine with halved gooseberries, sliced bananas, sprinkled with lemon juice. Add Strawberries, tossed with sugar, and gently toss together all the fruit. Chill about 25 minutes. Sprinkle with Cointreau and chill.
Just before serving, sprinkle with scant sugar and place in a to-the-table bowl,
Or place on crisp Bibb lettuce on individual salad dishes. Decorate with Mint sprigs from the garden.

STRAWBERRY CLOUDS

1 quart Strawberries - 2 T. sugar
1/4 cup Cherry Brandy liqueur
1 cup sour cream — 2 T. br. sugar

Prepare Strawberries. Combine
with sugar and liqueur. Chill
for 30 minutes. Stir gently and
refreeze. Combine sour cream
and brown sugar.

In an hour remove 1/2 c. berries
from syrup. Mash and stir into
sour cream / sugar. LayER berries
and sour cream. Drizzle with liqueur
(use sherbet glasses).

CRYSTALLIZED HERBS

After sugar, wine and vinegar,
Strawberries love herbs.
Not everyone is so lucky
to have herbs at their
door step, but why not try
some simple way to conserve
flowers, stems or leaves, to
use as a garnish for strawberries.
The recipe given, using mint
leaves (they will grow any-
where — well, at least in
the Northeast, midwest and
thereabouts), can be used
for any other edible plants.
Some suggestions that
lend themselves to Strawberries
are angelica, violet, lavender,
rosemary, marigold and rose
petals; lemon balm, violet and
bergamot leaves.
The principle goes back hundreds of years. Natural juices
are gradually replaced with sugar

Crystallized Herbs (continued)

by repeated immersion in a thick syrup.

MINT LEAVES

Gather mint leaves on a sunny morning after the dew has dried.

Beat an egg white with a fork until it is barely opaque. Hold each dry leaf by stalk and dip into egg until well coated. Now dip leaf into sugar, coating thoroughly.

Lay coated leaves in single layer on wax paper on a wire rack. Cover with another piece wax paper to protect from dust. Dry leaves in a very low oven with door ajar. When very dry + brittle, store between layers of wax paper in an airtight container.

STRAWBERRY-PINEAPPLE JAM

1 quart strawberries (2 c. prepared)
1 cup drained crushed pineapple
2 Tablespoons lemon juice
5 cups sugar
½ bottle liquid fruit pectin

Follow directions for Strawberry - Rhubarb jam on page 44.

(the barn at Cranberrie Cottage)

STRAWBERRY 'n RYE

One fifth Rye Whisky
One quart sliced Strawberries
One cup sugar

Combine in a large jug and
shake well - daily - for one
to two weeks.
Strain and serve - cold - as
a liqueur.

STRAWBERRY STUFFED PEACHES

Pare large peaches. Cut a
slice from Top. Remove pits
and do not break the fruit.
Fill the hollow with crushed
Strawberries sweetened to
taste with honey. Sprinkle
with sugar and nutmeg.
Pour your favourite custard
over top and bake

OR

Serve cold soft custard
with the uncooked chilled
fruits.

119

SUMMER PIE
(with ice cream)

Combine 1 cup mashed Straw-
berries with 12 marshmallows.
Heat slowly, stirring with a
wooden spoon until smooth.
Remove from stove and stir
gently until foamy in texture.
Cool well.
Beat 2 egg whites until stiff.
Add 1/4 cup sugar and dash
of salt.
Gently fold the two mixtures
together.

Prepare and bake a cereal flake
shell. (See page 172). Fill shell
with 2/3 qt. French Vanilla ice
cream. Cover with 1 cup
sliced STRAWBERRIES, topped
with the hot meringue.
In a very hot oven - 450°F -
brown quickly under broiler,
about 30 seconds. Remove
and tuck whole berries
into the meringue.

120

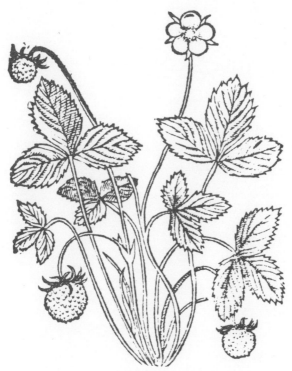

MARITIME AMBROSIA

3 grapefruits - large if available
1 pint halved strawberries
¼ cup shredded coconut
¼ cup honey

Remove peel and membrane
from each grapefruit. Save the
juice, add pulp and honey and
toss lightly. Add strawberries
and toss again. Before serving
add coconut and toss again.
Great for breakfast or dessert.

Nova Scotia Special
(Strawberry/Rhubarb Pie)

Prepare your favourite pastry for
a 9-inch double crust pie. Add
2 Tablespoons orange peel, grated
very, very fine. Preheat oven
to 450°F.
Combine and spoon into crust:
   1 pint strawberries, halved
   3 cups rhubarb, chopped
     or 1 pkg. (160gs.) frozen rhubarb
   1½ cup sugar — 5 Tablespoons flour
   1 tsp. each lemon and orange
               rind, grated
   2 Tablespoons butter, softened
         Salt

Cover with top crust. Sprinkle
sugar on top and bake for
10 minutes. Lower oven to
350°F and cook about 40 minutes
longer. Cool. Serve plain.

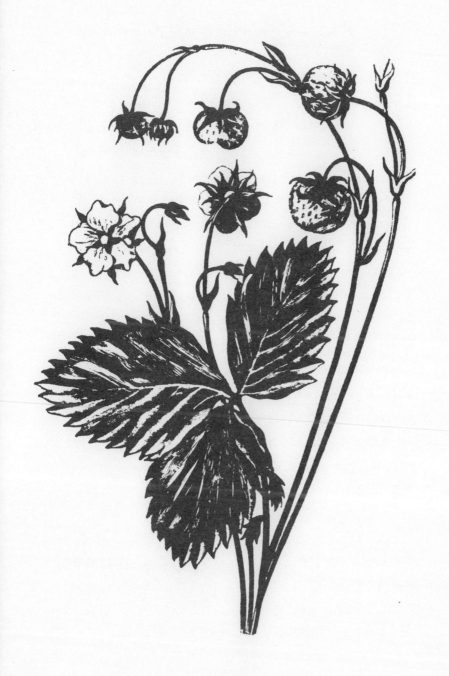

123

HAPPY VALLEY JAM

1 quart strawberries (2 cups)
1 3/4 pounds sugar (4 cups)
3/4 cup water - 1 box pectin
                    (1 3/4 ounces)

Crush berries, one layer at a
time. Add sugar, mix and let
stand 10 minutes. Dissolve the
pectin in water, bring to boil.
Stir and boil 1 minute. Now
add pectin to berries, stirring
constantly. Ladle into jars
but do not use paraffin. Let
stand at least 24 hours. Freeze

SUMMER SHERBET

4 quarts sliced Strawberries
4 cups sugar - 2 2/3 cups milk
2/3 cups lemonade (frozen)
1/2 teaspoon nutmeg

Mix berries and sugar. Let
stand about 2 hours. Purée.
Add milk, juice and seasoning.
Mix. Freeze in loaf pans, stirring
a few times before firm.

CORNWALLIS COOLER

1 large package frozen strawberries
1 cup fresh or frozen raspberries
3 Tablespoons honey — Rye Whiskey
1 each sliced lemon and orange
6 cups cold dry gingerale

Blend first three ingredients.
Before serving, chill for at
least 5 hours.
Combine with citrus slices
and cold gingerale.
Pour over ice block in bottom
of punch bowl. Add soda
water and Rye whiskey.

STRAWBERRY RICE

½ cup rice - 1½ pints milk
Orange-peel - ½ cup sugar
2 Tablespoons granulated gelatin
1 teaspoon Grand Marnier
½ cup cold water - 1 cup cream
Strawberries - dash of salt

Combine milk with thin strips
of orange rind (I use a potato
peeler) in top of double boiler.
When steaming, add the rice,
washed well, and a dash of
salt. When rice is tender and
moist, add Grand Marnier,
sugar and gelatin that has
been soaked in cold water.
Gently mix, and when it begins
to thicken, fold in the stiffly
beaten cream to which has been
added one cup of mashed
strawberries. Pour into a
mold and chill.
Serve for a special event, with
sweetened crushed straw-
berries. (Sometimes I use
only sugar, or combine it with
Grand Marnier)

126

"This berry is the wonder of all the fruits... one of the chiefest doctors of England was wont to say that God could have made, but God never did make, a better berry... The Indians bruise them and make strawberry bread."
Roger Williams
Rhode Island, 1643

STRAWBERRIES IN FONDANT

Fondant can be made ahead and stored in a covered stone crock, with a piece of damp cloth underneath the cover.

2 cups sugar - 1 cup water
1½ Tablespoons glycerin

Heat sugar and water over low heat. Stir often. When the sugar is dissolved, add glycerin and boil until a few drops in water form soft ball. Brush down sides of pan often with wet pastry brush so crystals will not form.
When syrup is ready, pour on lightly oiled marble or enamel top of table. When edges begin to stiffen, take a wide spatula and turn edges towards the centre. Continue until syrup is a creamy mass. Cover with a damp cloth and leave for 15 minutes. (continued)

STRAWBERRIES IN FONDANT

Place as much fondant as you need in a small saucepan. When it is runny enough to use as a dip, (very low burner) take each Strawberry by the stem, dip into fondant for a few seconds. (Fondant hardens quickly.) Remove and hold by the stem until fondant is firm, then place on wax paper.

Fondant can be varied by flavouring with your favourite liqueur.

BERRY HEAVEN

2 pints Strawberries
Confectioner's sugar
½ glass white dry wine
1 pint raspberries - 1 tsp. sugar
½ cup toasted slivered almonds

Place whole strawberries
in pretty serving bowl. Sprinkle
with sugar, add wine and
cover for 1-2 hours. Purée
raspberries adding 1 teaspoon
sugar. Pour over strawberries.
Top with almonds. Serve as
is or with sweet cream.

"Garden fruits are never or only
very rarely sold and anybody can
go into a garden and eat as
much as he likes without
restriction. Only strawberries
are sold although I saw them
growing wild for miles, they are
very expensive when cultivated."
               Alonso de Ovalle, Missionary
                    Chile, 1646

130

MARGAREE SHERBET

2 cups Strawberries
5 cups chopped rhubarb
2½ cups sugar, sweetened with
                    Rose. water

1 cup water

Combine rhubarb and water.
Cover, then simmer until the
fruit is soft. Mix well with the
sugar and pour into bowl.
Mash strawberries in blender
and add to rhubarb and mix
well. Freeze in two pans.
Put into a large bowl and beat
until smooth but do not melt.
Refreeze. Beat again and refreeze.
Soften for 30 minutes in the
refrigerator before serving.

1000 acres of Strawberries
employs 28,523 man days per
year. Labor to harvest 600
acres is 18,960 man days,

SPONGE CAKE

⅔ cups Sugar
1 T. strawberry
        syrup
4 egg yolks
4 egg whites, stiff
¾ cups sifted flour
3 T. hot butter (not melted)

Beat sugar, syrup
and yolks until thick.
Fold in whites, beaten
stiff, and flour. When well
mixed, add butter. Bake in
two round tins, lightly greased
and dusted with flour. Bake
about 50 minutes at 350°F.
P.S.
I've heard that the only real
recipe for sponge cake is
when you borrow all the
ingredients!

VIRGIL referred to the strawberry
as a "child of the soil," and the
"food of the Golden Age."

132

ST. CROIX SALAD

2 envelopes gelatin (dissolve
    by directions on label)
1/4 cup dry sherry
2 cups Strawberries
1½ cups Blueberries
Topping

Cook together until syrupy.
Arrange in a ring mold. Chill.
Top with a few whole strawberries and blueberries,
rolled in vanilla sugar.

In the Maritimes, the most popular cultivars are BOUNTY, MICMAC and, VEESTAR, from Ontario. In British Columbia the most predominant cultivar is TOTEM.

The cultural influence of the strawberry is evident from the names given to the cultivars throughout the world.

SUNSHINE STRAWBERRIES

So many people sent me a variation of this recipe that I include it although I've never made nor eaten sunshine strawberries - except those picked in the meadow of course!

Mix equal parts berries and sugar. Stand in cool spot for an hour, then place over burner until the sugar is dissolved. Stir often. Pour into platters in full sun. Cover with windowglass at an angle so condensation will not drip on berries. Protect from

SUNSHINE STRAWBERRIES
— continued

flies and bugs with cheesecloth.
Food should cook, thickening in
1 to 3 days. Stir each day a
few times and always bring
into house at night.
When mixture has consistency
of honey, pour into prepared
jars and process about 10 mins.

This is an old recipe - sunshine
and strawberries.

France, especially around
Paris, has always been
famous for strawberry
production.

Flavoured Sugar

Sugar is used in most
strawberry recipes. Whenever
used as a garnish or dip,
try flavouring with a sprig of
Rosemary, or a vanilla bean,
kept in a tightly lidded canister
of sugar

STRAWBERRY MALLOW

Combine, chill and whip vigorously,
1 cup small marshmallows and ½ pint
heavy cream. Fold in 1 cup Strawberries, halved, sweetened to
taste with flavoured sugar.
Chill well before serving.

CHOCOLATE DIPPED STRAWBERRY

8 ozs. sweet chocolate - Brandy
1 quart Strawberries with stems
½ cup cream, whipped

Melt chocolate in top of double-
boiler. Stir constantly - about 10
minutes. Stir into whipped cream.
Cook about 5 minutes. Stir in
2 Tablespoons of Brandy.
Place berries in large bowl.
In another bowl put sauce
for dipping.

The first commercial strawberry
nursery in the U.S. was in Long
Island, N.Y. in 1730.

136

STRAWBERRY CRISP

2 large cans Strawberry filling
1¼ cups granola-like cereal
½ cup melted butter

Place pie filling in a baking
dish. Toss ~~together~~ the cereal
and butter. Arrange over the
pie filling. Bake in preheated
350°F oven for 10-15 minutes.
Can be served warm or cold.

During a period of drought
the wild strawberry plant
is dormant.

137

A timely gift of strawberries in May, to a royal household in 916, resulted in a knighthood being bestowed on a man named Berry. His name was then changed to Fraise, and later it became Fraser. (On the present Fraser coat-of arms there are three fraises or stalked strawberries.) Some Frasers went to Scotland in the mid 1000's. It was a descendent, Amédée Francois Frézier, who brought the Chilean strawberry to Europe in 1714.

FRUITS and HONEY
Whole Strawberries -(bite size)
Canteloupe balls - green grapes
Blueberries - Peaches (quartered)

2 Tablespoons lemon juice
½ pint clear honey
candied ginger, diced
Combine the three sauce ingredients in a blender until well mixed. Gently mix with fruit. Chill and serve with whipped cream. Top with strawberries

138

STRAWBERRY NECTAR

4 qts berries, cleaned and crushed
3 cups Sugar - 2 whole nutmegs
1 Tablespoon cloves
Cinnamon sticks - about 2, 3" each.

Strain crushed berries through a sieve and measure. Add sugar (about 1 cup per 2 cups juice.) Add spices and simmer about 25 mins. (Tie spices in cloth). Do not burn. Remove spices, boil juice a minute, then pour into jars.

FROSTED STRAWBERRY MELON

Melon - Strawberry filling
Cream cheese - milk
Lettuce - Fruit mayonnaise

Carefully peel melon. Cut
slice from one end and
remove seeds. Fill with
Strawberry Filling (see below).
Refrigerate until center is
firm. Slightly soften
cheese with milk and frost
entire outside of the melon.
Serve in slices on crisp
lettuce, topping with Fruit
Mayonnaise.

STRAWBERRY FILLING
Beat together ½ lb. cream cheese,
¼ cup heavy cream, ¼ cup Kirsch and
½ cup mashed strawberries.
Sweeten strawberries with lemon
sugar (see page 84).
This filling can be used in other
recipes.

140

SOUFFLÉ PUDDING

Place a thick layer of Strawberries in the bottom of a greased baking dish. Pour over it a soufflé batter. Bake in moderate oven about 45 minutes at 350°F or until done. Serve hot.

In western Europe greenhouse strawberries are available for fall, winter and early spring consumption. The cost is prohibitive in Canada.

SOUFFLÉ BATTER

¼ c. butter — ¼ c. flour
1 c. milk — ¼ c. sugar
3 eggs, separated
1 tsp. Strawberry juice

Combine butter, flour, scalded milk and sugar. Make white sauce and add beaten egg yolks and Strawberry juice. Mix, then fold in stiffly beaten egg whites. Bake as directed in recipe at top of page.

141

"Mama's little baby loves shortnin' bread" - and shortnin' bread, or shortcake as we now say, seems to be synonymous with Strawberries. As the words of the song go, it is true that before the civil war, southern slaves were given shortnin' bread while the white folks in the big house, ate cake.

SWEET SHORTNIN' BREAD

1½ cups sifted flour - 1/2 cup milk
2½ teaspoons baking powder - salt
3/4 cup sugar - 2 eggs, beaten
whipped cream - Strawberries

Sift flour, salt and baking powder together. Cream sugar and 1/4 cup shortening until fluffy. Beat in eggs, then add dry ingredients and milk. Beat well. Pour into two greased cake pans. Bake at 350°F for 30 minutes. Remove and cool. Cover one layer with sweetened crushed berries. Cover with the other layer. Top with crushed and whole berries (toss whole ones in sugar). Serve with whipped cream.

STRAWBERRY FIELDS FOREVER

"The strawberry is the symbol of perfect righteousness."

Not only is the strawberry a religious symbol but it dates to a relatively early origin.

STRAWBERRY BREAD

2 cups crushed Strawberries
2 cups sugar – 3 eggs
1 cup salad oil – 1 T. lemon extract
2 cups flour – 1 cup Quick Oats
1 T. cinnamon – 1 tsp. baking soda
½ tsp. baking powder – 1 tsp. salt

Beat eggs and sugar, adding oil and lemon. Mix in flour, oats, cinnamon, soda, salt and baking powder. Fold in berries and mix gently but well. Pour into 2, 4x9 loaf tins that are greased and dusted with flour. Bake at 350° for 1 hour.

144

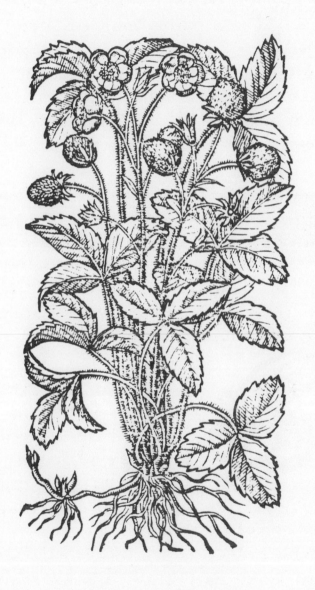

SEPTEMBER STRAWBERRY PIE

Prepare Graham Cracker Crust
(see directions on box)
1 10-ounce pkg. frozen strawberries
2 egg whites — 1 cup sugar
1 Tablespoon lemon juice
Dash of salt - 1 tsp. vanilla
1/2 cup cream -(whipped)

Combine strawberries, sugar, lemon juice, salt and egg whites. Beat until stiff.

Gently fold in whipped cream and vanilla. Pile into Graham cracker crust. Freeze

STRAWBERRY ANGEL

Take a cooled angel cake, and cover it, and fill hole in centre with mixture of whipped cream, sliced small strawberries and drambuie. Sugar to taste. Cover entire cake and chill. It's heavenly.

FRESH STRAWBERRY FROSTING

1½ cups mashed Strawberries
2 egg whites — 1½ cups Con-
        fectioner's Sugar
1 Tablespoon lemon juice

Combine in a large bowl,
the sugar, unbeaten egg
whites, lemon juice and Strawberries. Better if beaten
by hand. When peaks are
stiff, spread on cake.

CREAMY TOPPING
Combine ½ pint whipped cream
with 1 cup chopped marsh-
mallows. Chill at least ten
hours. Whip mixture, adding
sliced fresh strawberries. No
sugar is needed, but add a
few drops of your favourite
flavouring.

The 1984 strawberry crop in Nova
Scotia, exceeded 5 million quarts.
In 1983 it was 4.3 million quarts.

STRAWBERRY CONSERVE

1½ quarts Strawberries
1 cup seedless raisins
1 medium orange — Sugar
2 T. lemon juice

Mix fruits together, chopping coarse. Measure and add equal amount of sugar. Cook until mixture is thick, stirring constantly. Store in sterilized jelly jars.

This recipe makes a memorable gift for a new neighbor — or even an old neighbor!

RICH SHORTCAKE

2 cups flour - 1/4 cup sugar
1 egg - 4 tsp. baking powder
1/3 cup butter - 1 1/4 T. lard
1/3 cup milk - dash nutmeg + salt
          strawberries

Mix dry ingredients and then
cut in the lard. Add milk and
beaten egg. Bake in hot oven
about 12-15 mins, in a buttered
round layer-cake pan - or in
muffin tins. Split and cover
with cream sauce, then with
strawberries. Add top layer and
cover with strawberries that
are first rolled in Confectioner's
sugar. Spread again with
sauce and top with sliced
berries. Whip cream for top.
(see p. 164 for Cream Sauce recipe)

"The French called the Strawberry Fresas because of
the excellent sweetness
and odor." (1536)

DEVONSHIRE TART

Make your favourite pastry shell
to fit a 9" loose-bottomed tart
pan. Cool. Prepare 2-3 cups firm
strawberries. Beat together
until fluffy :
   1 (3oz) package cream cheese
   3 Tablespoons sour cream
Spread on bottom of shell and
chill. Mash enough of the
small berries to make 1 cup.
Force through sieve and add
enough water to make 1 cup.
Mix 1 cup sugar and 3 Tablespoons corn starch. Add ½ cup
water and sieved fruit. Cook
slowly until mixture has
thickened. Stir to cool.

Fill shell with remaining
larger berries, tips up. Pour
cooled cooked mixture over
top. Chill.   Garnish with
candied mint leaves and
violet blossoms.

CRÊME DESSERT

6 cups chopped rhubarb
4 cups crushed strawberries
3 cups sugar
2 cups whipping cream

Combine rhubarb and half of
the strawberries and cook
until barely tender. Keep the
mixture tart but sweeten a bit
with vanilla sugar. Whip the
cream and when peaks can
stand alone, fold fruits and
cream together. Pile into a
bowl and chill in refrigerator
for at least 8 hours.
Before serving, sweeten the
crushed remaining strawberries
and top the fruit mixture.
Garnish with Confectioner's
sugar and serve with hot
tea biscuits.

Throughout history there is ref-
erence to " strawberries and
cream" rolled in sugar, marinated
in wine."

151

BRANDIED FRUITS

Brandied fruits were recorded hundreds of years ago. Use glass or earthenware jugs. Old recipe books said to tie a piece of bladder over top to seal. Today most people use plastic. Cover one week, then seal with a lid. Insert a piece of paper between the brandy and lid so nothing touches the fruit or brandy. Use enough brandy to cover the fruit so it will not ferment.

Spread a little Happiness...

Strawberry Jam

Make Your Own Jams and Jellies

"Here also are Strawberries, I
have lien downe in one place in my
corne field and in the compass of
my reach have filled my belly in the
place."   Massachusetts, 1625

BASIC STRAWBERRY PUNCH

1 pkg. (16 ozs) frozen Strawberries
Combine thawed berries in large
punch bowl with:
   3 cups orange juice
   3/4 cup lemon juice
   1 large dry Gingerale - cold
   1 large can Pineapple Juice - cold
Mix well. Serve over crushed
ice in glasses, with a ring or
block of ice in punch bowl.

For a PARTY- PARTY, fill ice
cubes or ring with Strawberries.
Add 4 cups light RUM and
another 1/2 cup lemon juice - or -
add 4 cups white wine.

"I know of no one country yielding
without art or industry so manie
fruites...great fields and woods
abounding with Strawberries..."
          Virginia, 1615

153

STRAWBERRY GEMS

½ pint crushed strawberries
½ pound confectioner's sugar
2 egg whites, beaten

Beat eggs until very stiff. Fold in sugar and crushed berries. Drop from a spoon on a layer of waxed paper on a baking sheet. Bake in a very low oven, about 200°- 225°F.

The strawberry is a fast liver - it destroys itself quickly - high rate of metabolism!

A ripe strawberry could be compared to fruit of a rose (the hip, full of seeds) turned inside out.

"Long about knee-deep in June,
'Bout the time strawberries melts
On the vine ...
James W. Riley
Knee-Deep in June

154

Four strawberry varieties
out of Kentville, N.S. are now
grown in New York, Indiana,
Michigan, Ontario, Quebec,
New Brunswick and "probably
the U.S.S.R." These were
named: BOUNTY ('69), MICMAC ('71),
KENT ('81) and GLOOSCAP ('83).
ACADIA was developed in 1962.

STRAWBERRY ORANGE HONEY CONSERVE

2 cups crushed strawberries (1qt.)
1 medium orange – 1 cup honey
1/4 cup chopped walnuts
3/4 cup water – 3 cups sugar
1 box fruit pectin crystals

Follow directions as pc
Strawberry-Rhubarb jam
(page 44). Section orange,
discarding peel and membranes.
Dice orange and add nuts to
berries, then mix honey and
sugar into fruit mixture.
Makes 6 cups.

SUMMER CHICKEN SALAD

2 cups diced cooked chicken
1 pint halved strawberries
1 sliced banana   - lettuce
1 can pineapple chunks
1 T. lemon juice - ½ c. dressing
   Toasted slivered almonds

Sprinkle lemon juice over the banana. Add chicken, strawberries, and pineapple. Toss with slivered almonds and serve on a generous bed of crisp lettuce.

Tiny warm tea biscuits and white wine complete this summer luncheon salad.

Some historians feel that the tenderness found in 14th century paintings and prayer books, including sketches of strawberries, was reflection of the gentle Saint Francis of Assisi who loved all of God's creation.

"If a person loses respect for any form of life, he will lose his reverence for all of life." Albert Schweitzer

STRAWBERRY MILKSHAKE

Clean, hull and chop ½ cup fresh strawberries. Combine with 2 cups cold milk and 4 scoops ice cream. Blend at high speed for 5-7 minutes.

Experiment with different ice creams. Vanilla allows for a more authentic strawberry shake, but Banana or Ginger is also very good.

SUMMER ICE

Purée 2 quarts strawberries. Cool. Boil together 1 cup sugar and 1 cup water about 5 minutes. Cool. Combine and stir in strawberry juice. Freeze to a mush.
Makes about six glasses.

STRAWBERRY SOUP

1½ qts. fresh strawberries
1½ c. sugar — 6 T. lemon juice
3 c. water — 2 c. whipping cream

Combine all except cream.
Heat slowly until it comes to
a boil. Cook about 8 minutes,
then remove and cool. Blend
half the cream and half the
mixture at low speed. Now
add the remaining half and
repeat. Store soup in the
refrigerator in covered container for at least 4 hours.
Serve cold.

SUMMER SOUP

3 cups water   3 T. sugar
3 T. cold water   3 T. corn starch
Nutmeg   Whipped cream
2 cups fresh Strawberries

Add strawberries, sugar and
spice to boiling water. Mix
cornstarch in 3 Tablespoons
cold water, then add the
strawberry syrup that has
boiled ten minutes. Stir well.
Cook until clear and thick,
stirring constantly, about three
minutes.
Serve cold, with dabs of
whipped cream

"The annual harvest, (strawberries)
I am told, is always of such a
nature that it affords plenty of
bread for the inhabitants."

          Pehr Kalm, Swedish Botanist
          Raccoon, N. J. 1772

"Strawberries in abundance - one
may gather half a bushel in a
forenoon."   Massachusetts 1634

STRAWBERRY STREUSEL

Combine about 4 cups Strawberries with 1 cup Sugar, 1/4 cup
all-purpose flour and 1/2 teaspoon
nutmeg.
Pour into a baking dish that
can go to the table.
Combine:
  1 cup all purpose flour
  1/2 cup rolled oats
  1 cup brown sugar
  1 stick butter (melted but cool)
Sprinkle over the Strawberries.
Bake at 350°F for 40 minutes.
Serve cool, topped with
whipped cream mixed with
sliced strawberries.

A native Hawaiian Strawberry
was discovered 150 years ago
by a French horticulturist.
The discovery was accidental.
The fruit was called Ohelo,
meaning a pink-colored berry.

ST. Louis Strawberry

1 cup crushed Strawberries
1 pint heavy cream — sugar
1 T. Drambuie - pinch salt (small)

Whip and drain cream. Mix with
Strawberries, drained well and
tossed with sugar. Add Drambuie.
Pack into refrigerator trays,
for at least three hours. Serve
with sweetened sliced Strawberries topped with toasted
slivered almonds. Drizzle Drambuie.

161

Thos. Austin (1888), authority on early
English cookery, gives fine, old recipes,
dating to 15ᵗʰ century, for strawberry
puddings and the like:

"Take strawberys when in season, wash them in good red wine and press them through a cloth into a pot. Add almond milk [almonds soaked overnight and crushed in water to a fine consistency]. Add amyndoun [wheat soaked for several days in salt water, dried and crushed], and rye flour for thickening. Boil. Add dried currants, saffron [dried stigmas of *Crocus sativa,* L.], pepper, a good plenty of sugar, powdered ginger, cinnamon, galingale [powdered root of sedge, *Cyperus longus,* L.] Flavor with wine vinegar and suet. Color with alkanet [dyestuff made from roots of *Anchusa officinalis,* L.]. Sprinkle with pome garnad [pomegranate seed] and serve it forth."

BITTER SWEET PANCAKES
(Filling)

1 cup orange bitter marmalade
2 cups crushed strawberries
1 Tablespoon rum - Sugar

Make crêpes. Combine above.
Sweeten only if too tart - but
it should be more tart than
sweet.
Spread mixture on cooked
crêpes. Roll up crêpes and
serve on a bed of orange
slices that overlap to cover
the dish. Garnish with vanilla
sugar and shreds of orange
rind, cut very thin.
Orange Shreds: Remove about
four pieces of peel with potato
peeler. Using sharp knife, cut
tiny shreds. Put in cold water.
Bring to boil. Strain. Toss in
plastic bag with Confectioners
sugar.

"The leaves are an excellent pot herb
and the fruit a most wholesome berry."

163

STRAWBERRY CREAM SAUCE

1 egg (separated)
1 cup Confectioners' Sugar
1/2 cup heavy cream
1/4 cup milk
1 cup Strawberries

Beat egg white until stiff. Add yolk and beat. Add remaining ingredients. Beat slightly. Use on puddings and other desserts.

In 1802, the Austrian Scarlet Strawberry, from Nova Scotia, was obtained in Germany for the Royal Botanic Gardens at Kew.

BEAR RIVER COOLER

1/2 cup sliced Strawberries
1/4 cup honey - 1 cup Yogurt
1 cup chunk pineapple
3 T. sugar - 1 c. crushed ice.

Blend until smooth. Serve in tall chilled glasses.

(chill yogurt and pineapple and berries before blending)

More than half of the total world Strawberry production is from Europe. In 1977, for instance, some countries produced over 30 thousand tons:

Poland - 183.0   Germany 30.1
Italy - 152.0   (Fed.)
France - 73.9   Romania - 30.0
United Kingdom - 32.

FRUIT SALAD JAM

1 pound each peaches and pears
1 pint strawberries - 3 bananas
1½ cups chunk pineapple
6 cup sugar

Simmer together all except the bananas, until barely cooked. Add sliced bananas and sugar. When sugar is dissolved, boil quickly until jam will set when tested. Pour into sterilized jam jars and cover while hot.

Geese sometimes used to
control weeds in Strawberry
fields. Not a new idea, used
by cotton farmers and in
vineyards. But geese won't
eat lambsquarters & pig weed!

CREME FRAÎCHE

1 cup heavy cream
1 teaspoon commercial buttermilk

Pour into clean jar. Stand in
a warm place until thickened.
Stir well. Chill in refrigerator
24 hours before using.
This raw cream is a fine topping
for strawberries.

Some production figures for 1980:

U.S.A.	224 metric tons	
Japan	115	
Mexico	110	
Poland	100	
Canada	19	Ontario, Quebec, British Columbia and Nova Scotia

GLAZED STRAWBERRY PIE

1 quart Strawberries
½ cup sugar — ½ cup water
Scant ¼ c. corn starch
½ tsp. vanilla extract
¼ cup sliced toasted almonds

Reserve 1 cup berries. Purée.
Put whole berries into a baked
shell. (Use your favourite recipe.)
Place berries, the stems down,
and cover the bottom.
Combine sugar, water, cornstarch. Cook and stir over
medium heat until mixture
begins to boil. Add vanilla
and simmer for 5 minutes.
Cool. Pour evenly over berries
in pie shell. Sprinkle with
almonds. Chill 3-4 hours.
Garnish with whipped
cream, if desired.

strawberries — "delicacies of the
garden and the delights of the
palate." (1592)

The 1984 strawberry season is over — but not forgotten. It is the largest strawberry crop produced by Nova Scotia growers.

BLUENOSE STRAWBERRY SOUP

This recipe can be served as a soup or — as a dessert!

2 cups sliced strawberries
1 cup sour cream
1 cup half-and-half
1/4 cup sugar
2 Tablespoons brandy
1/2 teaspoon vanilla extract
(Extra sliced Strawberries and mint sprigs.)

Blend together all ingredients except the extras. Blend until smooth, about 30-45 seconds. Serve in chilled cups and garnish with sliced strawberries and mint sprigs.

HOLIDAY JAM

6 cups Strawberries
3 cups Raspberries
6 cups sugar — 2 T. lemon juice
1 tsp. grated orange rind
3 T. Grand Marnier

Combine and cook slowly.
Stir often — cooking about
30-45 minutes or until thick.
Too much cooking takes away
the flavor. Makes about 8,
6 oz. jars.
This recipe, made during berry
season, makes a special gift
throughout the year, especially
at Holiday time.

A favourite combination in the
early gardens of English royalty,
were "Strawberries, Primroses
and Violets."

169

SUNDAE SAUCE

Purée a 14-oz package of frozen
strawberries with 2 Tablespoons
Cointreau. Taste and sweeten if
you like — or add more liqueur.
Store in refrigerator in covered
container.
Serve over ice cream.

Do not grow raspberries, potatoes,
tomatoes, egg plants or peppers
near straw berries — or in the
same soil for at least four years.

STRAWBERRY PRESERVES

5 cups whole strawberries
(about 1½ qts. small berries)
5 cups sugar — ¼ cup lemon juice
½ bottle liquid fruit pectin

Measure 5 cups strawberries,
(pack well but do not crush)
in large saucepan. Add the
sugar and mix well. Place
over high heat and bring
to boil - stir gently so berries
will not crush. Let stand
about 5 hours.

Add lemon juice to fruit, bring
to rolling boil for 2 minutes.
Remove from heat and add
liquid pectin. Skim off foam
with metal spoon. Stir and
skim for 10 minutes. Ladle
quickly into jars. Cover at
once with ⅛ inch paraffin.

The strawberry grows underneath the nettle,
and holesome berryes thrive and ripen best
neighbour'd by fruit of baser quality.   *(Henry V, I, i)*

"A farmer returning home in his wagon, after delivering a load of (strawberries/produce), is a more certain sign of national prosperity than a nobleman riding in his chariot to the opera."
from <u>California Farmer</u>, 1856

CEREAL PASTRY

1½ cups crushed cereal
  (I especially like Grape Nut
                    Flakes)
¼ cup sugar ⌐ ½ cup butter

Mix crumbs and sugar. Stir in melted butter. Line pan by pressing firmly in place. Chill about 20 minutes, or bake in moderate oven 350°F about 10 minutes. Cool.

The Russians first added the new strawberry names to the scientific botanical literature, following a voyage around the world in 1816.
Kotzebue, 1821

172

CREAMY STRAWBERRY

2 cups crushed Strawberries
1 cup sliced Strawberries
½ lb. creamed cottage cheese
½ lb. cream cheese
2 cups cream

Combine cheese and salt
and slowly add the cream.
Beat constantly until thick
and smooth.
Turn one-half of mixture
into a mold. Add thick layer
of crushed berries. Top
with remainder of mixture.
Chill well for 4-6 hours, at
least. Serve with sliced
strawberries sweetened to
taste. Garnish with sprigs
of fresh or candied mint.

strawberries, unlike
other domesticated
fruits, do not improve
in flavour with
cultivation.

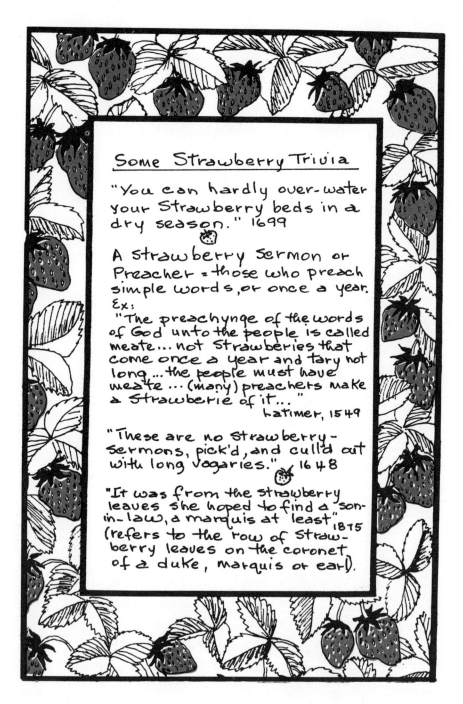

Some Strawberry Trivia

"You can hardly over-water your Strawberry beds in a dry season." 1699

A Strawberry Sermon or Preacher = those who preach simple words, or once a year.
Ex:
"The preachynge of the words of God unto the people is called meate... not Strawberies that come once a year and tary not long ... the people must have meate ... (many) preachers make a Strawberie of it..."
                    Latimer, 1549

"These are no Strawberry-Sermons, pick'd, and cull'd out with long vagaries." 1648

"It was from the strawberry leaves she hoped to find a son-in-law, a marquis at least" 1875 (refers to the row of straw-berry leaves on the coronet of a duke, marquis or earl).

SUMMER SALAD

1 quart Strawberries
Romaine or Bibb lettuce, torn
1/2 cup mayonnaise - 1/4 cup sugar
1/2 cup sour cream -
1/4 cup each poppy + sesame seeds
1/2 tsp. lemon juice
1/4 tsp. ground ginger

Drain washed berries on paper towel. Slice. Combine everything else except the lettuce. Serve on individual plates. First, arrange the sliced strawberries on the greens. Drizzle the SUMMER SALAD DRESSING over the top. Garnish with fresh red rose petals (wild roses or moss roses have the most flavour). As noted throughout, my preference is to use home-made mayonnaise.

In Rhode Island the Narragansett Indians called the strawberry WUTTAKIMNEASH, meaning "heart berry."

STRAWBERRY CHIFFON

Prepare an Almond Crumb Crust.

1¾ cups sliced Strawberries and a handful of whole berries. Cover sliced fruit with ½ cup Sugar. Put it aside for 20 mins. Sprinkle 1 envelope gelatin over ¾ cup cold water and 1 T. lemon juice, until softened, about 6 minutes. Cook slowly adding pinch of salt. Beat until mixture thickens, then fill shell with sliced berries, cover with thickened gelatin and lemon. Top with whipped egg whites. (Some versions of this recipe combine the gelatin mixture with egg whites and beat until stiff.) Garnish with sliced Strawberries.

Strawberries are sometimes protected from birds by netting, during the most susceptible period. Firecrackers are also used but unfortunately, the Robins love firecrackers!

In Oregon, berries are stemmed by hand-pickers and put in molded plastic containers and flats. Pickers are paid by the flat.

SPRING MUFFINS

¾ cup sliced Strawberries
½ cup Rhubarb - chopped fine
1¾ cup all-purpose flour — 1 egg
¾ cup sugar — 2½ tsp. baking pwdr.
¾ cup milk — ⅓ cup vegetable oil
6 medium Strawberries, halved - salt

Preheat oven to 400°F. Mix together flour, ½ cup sugar, salt and baking powder. Then combine egg, slightly beaten, milk and oil. Stir into first mixture. Gently fold fruit into the batter. Fill well greased and floured tins ⅔rds full. Place a half Strawberry on top of each muffin and sprinkle with sugar. Bake about 30 minutes at 400°F.

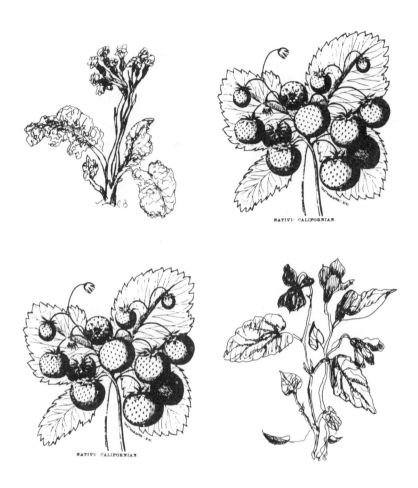

NATIVE CALIFORNIAN.

NATIVE CALIFORNIAN.

The garden accounts of Henry VIII (1539)
showed payment made for
   "gatheryng of 34 bushells of stewberry
   rot (post harvest trimmings), primrose,
   and violet of 3d the bushell."
Strawberries - Primroses - Violets

TURKEY IN STRAWBERRY LEAVES

Gather strawberry leaves, press them in the distillery until the aromatick perfume thereof becomes sensible. Take a fat turkey and pluck him carefully and baste him, then enfold him in the Strawberry leaves.
Boil him in water from the well, and add rosemary, lavender, thistles, stinging nettles and other sweet-smelling herbs.
Add also a pinte of canary wine, a halfe a pound of butter and one of ginger passed through a sieve. Serve with plums and stewed raisins and a little salt. Cover him with a silver dish cover.
                An old recipe.

"Strawberry leaves have a most cordial smell, next in sweetness to the muskrose and violet."
                                Francis Bacon

ASPIC SURPRISE

2 pkgs. unflavoured gelatin
  (dissolve as directed on envelope)
2 cups canned apple juice
1 tsp. sugar — ½ tsp. salt
Dash of powdered cloves
1 Tablespoon cider vinegar
½ cup avocado slices, halved
½ cup sliced strawberries

Combine apple juice and
dissolved gelatin. Add sugar,
salt, cloves and cider vinegar.
When well mixed, chill until
somewhat thickened.
Fold in the fruit slices. Pour
into a ring mold and chill
well until firmly set.
Serve on bed of crisp greens
with STRAWBERRY HONEY
topping in centre. (see page 52)

Note: You may wish to sweeten
the strawberries.

STRAWBERRY MAYONNAISE

1 cup prepared mayonnaise
(or home-made — that's the best)

3 Tablespoons Strawberry syrup
1/2 cup Confectioner's sugar
1 cup heavy cream, whipped

Combine, mix well and Chill.

WHIPPED MAYONNAISE

1 cup mayonnaise
4 tsp. Confectioner's Sugar
1 cup heavy cream, whipped

Combine and mix thoroughly.

"Raw crayme undecocted, eaten with strawberries... is a rurall mannes bankit. I
have knowen such bankettes hath put
men in jeoperdy of theyr lyves." (1542)

> If frost do continue take this for a lawe
> The strawberries look to be covered with straw
> Laid overly trim, upon crotchis and bows
> And after uncovered as weather allows.

STRAWBERRY SANDWICH

To make 4 sandwiches, use 8 slices of bread - French or Italian - cut thin.
Fill sandwich with strawberry jam - about 2 Tablespoons each.
Spread outside with butter.
Brown each side quickly.
Sprinkle with confectioner's sugar.

Strawberries require small labor, but by dilligence of the gardener becommeth so great, that the same yieldeth faire and big Berries. . . . Eaten with creame and sugar they are accounted a great refreshing to men, but more commended being with wine and sugar.

Thomas Hill
The Gardeners Labyrinth, (1577)

Again, in The Profitable Arte of Gardening (1593), Hill wrote:

Strawberries be much eaten at all mens tables in summer for the pleasantness of them, which for a more delight in eating they dress with wine and sugar. Set in gardens the plantes will growe into the bigness of a mulberry if the earth before in beds be well drysed and dilligently tended of the gardner.

Whipped Strawberry Pie

Wash and hull 1 quart Strawberries. Pick out 1 cup of the poorest ones and add these to 1 cup of Sugar and 1 cup cold water. Boil for about 15 minutes. Strain, crushing berries with wooden spoon. Discard pulp.

Dissolve 3 Tablespoons cornstarch in ⅓ cup cold water. Add this to Strawberry juice, stirring constantly until mixture boils. Simmer until it is a thick syrup. Cut remaining berries in quarters. Cover them with the hot syrup. Chill well.

In a baked 9" pastry shell, spread a half pint of cream stiffly whipped. Top with chilled Strawberry mixture. Dot with whipped cream.

Linnaeus, the Swedish botanist, prescribed for himself a diet of strawberries and he was cured of the gout — or so he wrote.

Prior to 1766 no Europeans seemed aware of the importance of sex in strawberries. Female plants must be fertilized with Pollen from male plants.

"Thank you for the World so sweet."

Strawberries owe their adaptability to the "fortunate combination of a rare genetic makeup and hybrid origin from wild types of Canada, Virginia, Chile and California."

Stephen Wilhelm, Phd.
American Scientist
Dec. 1981

184

STRAWBERRIES in CHAMPAGNE JELLY

2 pkgs. unflavoured gelatin
½ c. water — ½ c. sugar
½ c. Strawberry juice
1 quart STRAWBERRIES
1 bottle pink Champagne (3¼ cups)

Soften gel in water. Heat until dissolved. Cool.
Dissolve sugar in Strawberry juice. Add gelatin mixture and champagne.
Pour into 6-cup mold. Chill until thickened. Then fold the berries, one by one, into the mixture. Chill about 5 hours. Serve garnished with sliced strawberries sprinkled with vanilla sugar.

Strawberries are found all over the world – "from Kashmir to Kamchatka, to Spain, Oregon + Hudson's Bay – in deep valleys or alpine heights."

STRAWBERRY JELLY
(FREEZER)

5 cups Strawberry juice
   (about 2½ quarts berries)
6 cups sugar — ¾ cup water
1 package powdered pectin

Crush berries. Drip through damp jelly bag. Add sugar, stir and let it stand for 10 minutes.
Now dissolve pectin in the water. Stir and boil 1 minute. Remove from heat and mix with juice. Pour into prepared jars leaving ½ inch at the top. Stand about 24 hours, then freeze.

"The strawberry plant is called Fragaria because the fruit was called Fraga by Virgil."

Leonhart Fuchr. 1543

GRANVILLE COOKIES

1 cup all-purpose flour
2/3 cup Confectioner's sugar
7/8 cup butter
½ cup corn starch
½ tsp. baking powder
Strawberry Jam

Cream sugar and butter.
Sift together flour, baking
powder and corn starch.
Roll into small balls and
place on cold greased
cookie sheets. Press each
one with fork, or, using a
cookie press, pipe dough
to make a smooth cookie.
bake at 350°F for 20 minutes.
When cool, sandwich together
with Strawberry jam.
The jam is especially good
if mixed with a small
amount of whipped cream,
and grated lemon.

In mid 1700's the botanist became an
experimenter, trying to understand the
evolution of things in nature.

"wild or voluntary strawberries were not so good as those that are manured in gardens."

Tobias Venner, 1638

STRAWBERRY FILLED COOKIES

2½ cups sifted flour - 1 cup sugar
2 eggs, beaten — ¼ tsp. salt
2 tsp. baking powder — ½ tsp. lemon
½ cup shortening — 1 T. milk
      strawberry jam

Sift together flour, salt and baking powder. Cream sugar and shortening, then add eggs, lemon, the sifted ingredients and milk. Roll out on floured board and cut into circles, large enough to fold over and hold filling, or small enough to use two per cookie. Fill with jam. Sprinkle with sugar and bake in moderate oven (375°F) about 12 minutes. Makes 2½ dozen cookies. Variation: Use brown or maple flavoured sugar, and in filling, fold in nutmeats and/or liqueur.

OBLATE GLOBOSE GLOBOSE CONIC

CONIC SHORT WEDGE

LONG CONIC NECKED LONG WEDGE

Shapes of strawberries.

189

FRUIT SOUP

⅓ cup seedless raisins
¾ cup dried apricots
6 cups water
2 cups peeled sliced firm apples
½ cup sugar
¼ cup tapioca (quick-cooking)
4 cloves - 1 stick cinnamon (3")
4 lemon slices (thin)
4 orange slices (thin)
Dash of salt
1 cup Strawberry syrup
¼ cup blanched slivered
                almonds

Cover pot and simmer dried
fruits in water for 30 minutes.
Add apples, tapioca, sugar,
spices, citrus and salt. Cover
and simmer gently about 30 minutes.
Add strawberry syrup. Sugar
to taste.
Chill overnight. Sprinkle
with almonds when serving.

The Spanish explorers found Strawberries on sand dunes of California
in December, 1602.

Commercial strawberry production began in the United States in 1800, but with the same species until 1840. Since 1858 there has been an endless variety of strawberries available.

BOOTHBAY'S BEST

1 pint halved strawberries
2 cups chopped rhubarb
1⅓ cup sugar - 2 T. corn starch
Milk - Nutmeg

Prepare pastry for a 9-inch crust. Combine fruit and sugar. Dissolve corn starch in cold water and add. When well coated, fill crust with fruit mixture, dot with soft butter and sprinkle with ⅛ teaspoon grated nutmeg. Flute edges.
Make lattice top of pastry strips. Sprinkle with milk and sugar. Preheat oven to 400°F and bake 1 hour.

For market gardeners, strawberry sexuality had more than academic interest – it meant their livelihood.

SPRING YOGURT

1½ cups frozen unsweetened
               Strawberries
2 cups frozen pre sweetened
       Rhubarb sauce
1 cup plain Yogurt

Blend thawed fruit until smooth.
Add Yogurt. Beat until smooth.
Chill in individual Parfait
glasses. Garnish with
a bit of fruit juice saved
from thawed berries. Sweeten
to taste.

STRAWBERRY-LEMON FILLING

About 6 to 8 hours before serving, top each meringue with this mixture – or use nests.

Combine in top of double boiler and simmer for about 8 minutes:
- 1/3 c. sugar — 4 egg yolks
- 3 T. lemon juice — pinch of salt
- 2 T. finely grated lemon peel

When mixture has cooled, fold in 1/2 pint whipped cream, beaten to soft peaks, and about 3/4 cup sliced strawberries.

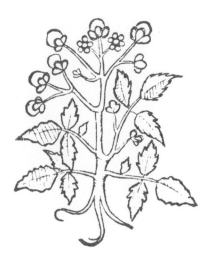

ffragaria   erperkrut

Frag ria eft frigide coplexioms. Valet contra
apoftem.a gutturis ifto modo· R·fucci fragarie
et aque pla̅ta̅ginis a̅n̅·᷈·iiij· mellis rofari· ᷈·j· fuc
ci moror̄ celfi·᷈·ʃ· albi grea balaftie a̅n̅·᷈·j· fiat
inde cu̅ parū aceri gargarifmus· Jtem aqua deco
ctioms fragarie valet cōtra fudoꝛem et caloꝛem
Et fi im tali decoctione fragarie diffolueref᷈ dra͛
gantū valet cōtra fitim· Cōfert ena vmu̅ deco᷈
ctioms fragarie et femis petrofilim ꝛfaxifrage

    First printed figure of the strawberry, from the *Herbarius Latinus Moguntiae* (Latin Herbal of Mainz) of 1484, giving the botanical name, *Fragaria*, and common name, *erperkrut*. The text is a compendium of the pharmaceutical properties of strawberries. The portion shown reads: "The strawberry is cooling to the skin. It helps against throat infections if used as a gargling compound formulated as follows: Strawberry juice and plantain water, 4 parts each, rose honey, mulberry juice, album grecum, balastien, 1 part each, mixed with some vinegar. Juice decocted from strawberries is good against perspiration and fever; if dragantum is added, it quenches thirst." The remainder of the text, not shown here, treats the uses of strawberries for other ailments, such as kidney stones, halitosis, bone fractures, bruises, wounds, and certain internal injuries.

VARIATION - ON A SOUFFLÉ

Make a plain soufflé, but
omit the flavouring. Add
1/4 cup chopped candied fruit
that has soaked overnight in
1/4 cup brandy. Add to batter
with the sugar.
When soufflé has baked
25 minutes, remove it and for
next 5 minutes cook with 1 cup
halved strawberries on top.
Sprinkle with sugar.

VARIATION - ON A FLAN

Cover bottom of flan with:
3/4 cup cream, whipped, and 1 T.
sugar and 2 T. Drambuie. Top
this with 2 cups whole strawberries.
Now put 1/2 cup currant jelly
in pan. Bring to a boil. Add
1 teaspoon cornstarch blended
with water and boil 1-3 minutes.
Add 1 teaspoon lemon juice and
1 T. Drambuie. Glaze the berries
immediately.

Wisconsin Strawberry Pie

Bake your favourite pastry
shell.
Mash enough berries to make
one cup. Mix together one cup
sugar and 2 T. cornstarch in a
2-quart pan. Gradually stir
in mashed berries and ½ cup
water. Cook over medium
heat and stir constantly until
mixture thickens and boils.
Boil about one minute - stirring
all the time. Cool.
Soften one package (3 ozs) cream
cheese - beat until smooth.
Spread on bottom of shell. Fill
with about 5 cups whole berries.
Top with cooked mixture. Chill.

STRAWBERRY/RASPBERRY JAM

1 quart strawberries
1 quart raspberries
7 cups sugar
½ bottle liquid
    fruit Pectin

Thoroughly crush raspberries,
one layer at a time. Measure
2 cups. Repeat with strawberries
and then combine to make 4
cups of fruit.

Add sugar to the fruit and
mix well. Place over high
heat, bring to a rolling boil
and boil hard 1 minute. Stir
constantly. Remove from
heat and stir in liquid pectin.
Skim off foam with metal
spoon. Stir and skim about
5 minutes. Cool slightly.
Ladle quickly into jars.
Cover immediately with
hot paraffin - about ⅛-inch.
Makes about 6½ cups.

CANADIAN SANGRIA

4/5 quart red Burgundy wine
2 ozs. Strawberry liqueur
3 ozs. Cognac
2 ozs. Triple Sec
2 ozs. light rum
2 ozs. peach brandy
1 lime, lemon and orange -
        thinly sliced and seeded
Black cherries with long stems
12 large Strawberries, halved
½ fresh pineapple, in chunks
1 large Club Soda - crushed ice

Combine wine and liqueur in large pitcher. Add fruit and chill at least 1 hour to blend flavours. Add chilled soda to taste, over crushed ice, just before serving. Spoon a small amount of fruit into each tall serving goblet and fill up with the Sangria.

Very Berry Best

Yukon Pie

Use a greased round pie pan.
2-3 cups halved STRAWBERRIES
4 egg whites — ¼ tsp. cream tartar
¼ tsp. salt — 1 c. sugar — ½ tsp. vanilla
whipped cream

Combine egg whites, salt and cream of tartar. Beat until stiff, gradually adding sugar and flavouring.

Spoon into pan. Bake about 15 minutes at 275°F. Turn off oven, leaving pan in overnight.

Before serving time, spread meringue with whipped cream. Cover with STRAWBERRIES. Top with another layer of cream. Garnish with a few whole strawberries.

199

STRAWBERRY SHORTCAKE

3 cups crushed Strawberries
2 cups sifted flour — 1/2 tsp. salt
4 tsp. baking powder — 1T. sugar
1/3 cup shortening — 3/4 cup milk
Butter — (Sliced bananas)

Mix and sift dry ingredients.
Knife in shortening. Add milk
slowly and when dough is soft,
Knead it on a floured board.
Cut with floured biscuit
cutter. Brush tops with
butter and bake at 450°F
for about 15 minutes.
While still hot, split biscuits
and butter both sides.
Fill the shortcakes with
heaps of crushed berries
sweetened to taste.
Top with fruit and then
whipped cream.
Some times shortcakes are
served with bananas too.
Personally, I prefer that
Strawberries only be used.

STRAWBERRIES with ROSEMARY

3 cups puréed Strawberries
2 T. fresh Rosemary blossoms
1 T. vanilla sugar
2/3 cup heavy sweet cream

Crush the Rosemary blossoms in wooden bowl. Add sugar and mix well. Combine cream and puréed strawberries. Fold in crushed blossoms. Chill well before serving.

Blossoms should be "pounded" with the sugar to preserve flavour, scent and goodness. Some flowers have been stored seven years.

Keep a Rosemary plant in kitchen all year round!

201

AN ANCIENT CAKE
(with strawberries
and Rose water)

Take halfe a pecke of flowre,
half a pinte of rose water, a
pint of ale yeast, a pint of
creame. A pound and a half
of butter, six eggs (leaves
out whites) four pounds
strawberries, one half
pound sugar, one nutmeg
(grated) and a little salt.

Work it very well and let
it stand half an hour by the
fire and then work it again
and then make it up and
let it stand an hour and a
halfe in the oven; let not
your oven be too hot.

This recipe was sent to me by a
woman in Kalamazoo, Michigan

202

VICTORY STRAWBERRY PIES
During WWII, this was popular.

⅓ cup cool potato water
½ cake yeast - ¾ cup sugar
⅓ cup riced potatoes
2 T. lemon juice ~ 3 eggs
⅓ cup shortening, melted
1 cup sifted flour
1½ cups Strawberries

Combine potato water, crumbled yeast, cooled potatoes and ¼ cup sugar. Let rise 1 hour.
Add shortening, ¼ c. sugar, 1 egg (beaten) and flour. This should make a stiff dough. Knead well. Let rise until double in bulk. Roll out in 2 circles about ½" thick. Place in two greased pie pans - deep ones. Add Strawberries. Beat 2 eggs with lemon juice and rest of sugar and pour over berries. Sprinkle with nutmeg. Let rise, then bake 30 mins at 350°F.

ANN ARBOR Soup

1 cup strawberries, mashed
1 small cucumber - peanuts
Sour cream

Mash strawberries and mix with grated cucumber and very finely chopped handful of peanuts. Blend together, adding ½ cup sour cream, and later serving with a few sliced berries on top. Chill well.

This is not my favourite recipe but some people swear by it on a hot summer day. It can be thinned by adding water as desired.

WINTER MUFFINS

⅓ cup Strawberry preserves
2 cups all-purpose flour
¼ cup Sugar — 1 cup milk
1 T. baking powder — pinch of salt
1 egg — 3 T. butter
2 tsp. grated lemon rind

Sift together dry ingredients.
In another bowl combine egg,
milk, melted butter and lemon
rind. Beat gently, then add
to dry mixture.

Grease and flour muffin tins.
Fill them about ⅛ʳᵈ. Top with
one teaspoon preserves (or jam).
Fill tins with batter, sprinkling
sugar on top.

Bake until golden, about 30
minutes at 400°F. Serve warm.
Recipe makes about 12 (2½"
tins) muffins.
Variation: Mix 1 cup regular flour
and 1 cup Graham flour.

NUTTY SHORTCAKE

1½ cups unsifted flour
½ tsp. salt - ¼ tsp. baking soda
3 tsp. baking powder
½ cup dark brown sugar
⅓ cup shortening
½ cup chopped pecans or walnuts
1 egg   -   ¾ cup milk
1 quart strawberries,
        sweetened to taste

Grease and flour a round
8 x 1½" layer pan. Preheat oven
to 375°F.
Mix flour, salt, soda and baking
powder. Cut in the brown
sugar and shortening. Add the
nuts. Combine egg and milk
and stir into flour mixture.
Spread in pan and bake about
20-25 minutes.
Split into two layers. Fill
and top with strawberries.
Garnish with whipped
cream to which some juice
from berries has been added.

Virtue rejoice, tho' heaven may frown awhile;
That frown is but an earnest of a smile.
One day of tears presages years of joy.
For sufferings only mend us, not destroy.
Who feels the lashes of an adverse hour,
Finds them but means to waft him into power.
As health to bodies, bitter draughts impart,
So trials are but physic to the heart.
Sarah Wilkinson finished this work in 1819
Aged 11

The strawberry continued to have special meaning into 19th century.

Note strawberries on this silk-on-wool sampler from Ontario.

207

ROYAL RHUBARB/STRAWBERRY

1 c. sugar — 2 T. quick Tapioca
¼ tsp. salt — ¼ tsp. nutmeg
¼ c. orange juice — 3 c. rhubarb
1 c. sliced Strawberries. Butter

Combine all except berries and
butter in a plain pastry uncooked
shell to which is added 1 T. of
grated orange rind. Top with
berries and dot with butter.
Top with pastry.
Bake 10 mins. at 450°F, then
30 mins. at 350°F.

STRAWBERRY MERINGUES

4 egg whites — 1 cup Sugar
1/4 tsp. cream of tartar — pinch salt
1/2 tsp. currant jelly (for color)
1 1/4 cup whipping cream — Strawberries

Whip egg whites until stiff. Add
Sugar slowly, beating continuously.
Fold in cream of tartar and black
currant jelly. Drop by teaspoons-
ful onto prepared sheets. Place
on bottom shelf of oven, set as
low as possible. Dry for 2 hours.
When cool, join two together
with whipped cream into
which crushed, drained strawberries have been mixed. (You
can use Strawberry jam or
preserves, out of season)

Variation for serving: Remove
while warm, taking out soft
centre from underneath. Cool.
Just before serving fill centre
with whole strawberry or
crush berries and top with
whipped cream.

209

The challenge to strawberry production in the 80's is in the area of harvesting.

WHIPPED FRUIT DRESSING

2/3 cup sugar — 2 T. flour
2 eggs, beaten — 2 T. salad oil
3 T. lemon juice — 1 cup Strawberry juice
4 T. pineapple juice
1/2 cup heavy cream, whipped

Mix sugar and flour. Add remaining ingredients except cream and cook over low heat until thickened, stirring constantly. When cold, fold in whipped cream. Makes 2 cups.

Most strawberry production in Eastern Canada is pick-your-own.
Nova Scotia - 70%
Ontario - 80%
British Columbia- the focus is on processing - 90%

211

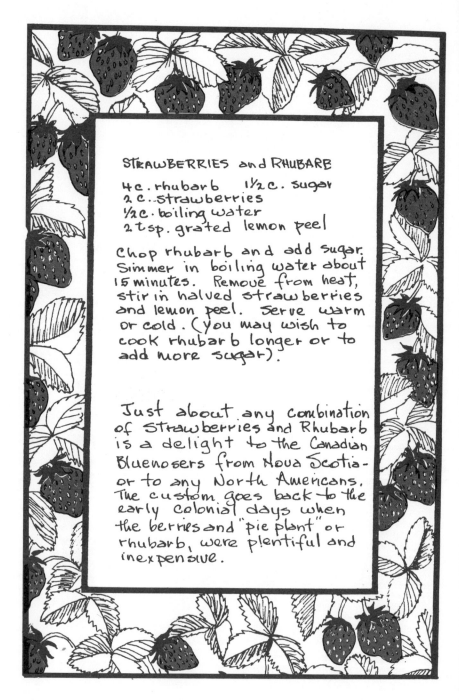

STRAWBERRIES and RHUBARB

4 c. rhubarb   1½ c. sugar
2 c. strawberries
½ c. boiling water
2 tsp. grated lemon peel

Chop rhubarb and add sugar.
Simmer in boiling water about
15 minutes. Remove from heat,
stir in halved strawberries
and lemon peel. Serve warm
or cold. (You may wish to
cook rhubarb longer or to
add more sugar).

Just about any combination
of Strawberries and Rhubarb
is a delight to the Canadian
Bluenosers from Nova Scotia-
or to any North Americans.
The custom goes back to the
early colonial days when
the berries and "pie plant" or
rhubarb, were plentiful and
inexpensive.

References

Bailey, Liberty Hyde, **American Naturalist,** 28:293, N.Y. 1894.
_____ , **Standard Cyclopedia of Horticulture,** 3:1272, N.Y. 1915
Barry, P., **The Fruit Garden,** N.Y. 1896 (revised edition)
Baxter, James P., **A Memoir of Jacques Cartier ... Voyages to the St. Lawrence,** N.Y. 1906
Bunyard, E.A., **The History of the Development of the Strawberry,** J. of Royal Horticultural Society, 39:550, 1913
Childers, Norman F., Ed., **The Strawberry - Cultivars to Marketing,** Proceedings of Nat. Strawberry Conf., St. Louis, 1980
Fletcher, S.W., **The Strawberry in North America,** N.Y. 1917
Hedrick, U.P., Ed., **Sturtevants' Edible Plants of the World,** Dover, 1972 (original 1919)
_____ , **The Small Fruits of New York,** Albany, 1925
Horticultural Science, Amer. Soc. for Hort. Science, Vol. 16, #6, December 1981
Mawe, Thos. & Abercrombie, John, **The Complete Gardener,** London, 1829
Sturtevant, E.L., "Whence Came Cultivated Strawberries?" **Trans. Massachusetts Horticultural Society,** 200, 201, 1888
Swartz, H.J., **The Strawberry,** University of Maryland, 1980
Wickson, Edw.J., **The California Fruits and How to Grow Them,** San Francisco, 1921 (9th edition)
Wilhelm, Stephen, "The Garden Strawberry: A Study of Its Origin", **American Scientist,** 264-271, Vol. 62
_____ & Sagen, James E., **The History of the Strawberry, From Ancient Gardens to Modern Markets,** Div. of Argicultural and Natural Resources, Univ. of California Press, 1974
Wright, John, **The Fruit Grower's Guide,** Vols. I-VI, London, 1892
American Federal and State and Canadian Federal and Provincial Departments of Agriculture - brochures, recipes, statistics, Etcetera.
Various University Departments of Pomology, especially Cornell U., Ithaca, N.Y.

> *"When I hear people say they have not found the world, or life so interesting as to be in love with it I am apt to think they have never . . . seen with clear vision the world they think so meanly of, nor anything in it, not even a blade of grass."*
>
> —W. H. HUDSON